Drying of Herbs, Spices, and Medicinal Plants

Drying is a key operation in processing of many plant-based foods and medicines for the purpose of preservation and retention of key attributes and active compounds. Therefore, it is essential to select suitable drying techniques to ensure a product is processed under optimal operating conditions. *Drying of Herbs, Spices, and Medicinal Plants* presents processing aspects of these three major global agricultural commodities. It offers an insight into the drying and product quality of herbs, spices, and medicinal plants, such as drying characteristics, equipment selection, physiochemical analyses, quality improvement, product development, storage, and shelf life as well as future developments.

- Offers the latest information on drying and processing technologies, research, and development
- Summarizes various drying techniques, their advantages and limitations, industrial applications, and simple design methods
- Presents guidelines for dryer selection
- Links theory and practice
- Envisages future trends and demands

Featuring chapters from expert authors in both industry and academia, this book is an important resource for those working in the chemical, food processing, pharma, and biotech industries, especially those focused on the drying of plants for food and medicinal applications.

Advances in Drying Science and Technology

Series Editor:
Arun S. Mujumdar, McGill University, Quebec, Canada

Drying and Roasting of Cocoa and Coffee
Ching Lik Hii and Flavio Meira Borem

Heat and Mass Transfer in Drying of Porous Media
Peng Xu, Agus P. Sasmito, and Arun S. Mujumdar

Freeze Drying of Pharmaceutical Products
Davide Fissore, Roberto Pisano, and Antonello Barresi

Frontiers in Spray Drying
Nan Fu, Jie Xiao, Meng Wai Woo, and Xiao Dong Chen

Drying in the Dairy Industry
Cécile Le Floch-Fouere, Pierre Schuck, Gaëlle Tanguy, Luca Lanotte, and Romain Jeantet

Spray Drying Encapsulation of Bioactive Materials
Seid Mahdi Jafari and Ali Rashidinejad

Flame Spray Drying: Equipment, Mechanism, and Perspectives
Mariia Sobulska and Ireneusz Zbicinski

Advanced Micro-Level Experimental Techniques for Food Drying and Processing Applications
Azharul Karim, Sabrina Fawzia, and Mohammad Mahbubur Rahman

Mass Transfer Driven Evaporation of Capillary Porous Media
Rui Wu and Marc Prat

Particulate Drying: Techniques and Industry Applications
Sachin Vinayak Jangam, Chung-Lim Law, and Shivanand S. Shirkole

Drying and Valorisation of Food Processing Waste
Chien Hwa Chong, Rafeah Wahi, Chee Ming Choo, Shee Jia Chew, and Mackingsley Kushan Dassanayake

Drying of Herbs, Spices, and Medicinal Plants
Ching Lik Hii and Shivanand S. Shirkole

For more information about this series, please visit: www.routledge.com/Advances-in-Drying-Science-and-Technology/book-series/CRCADVSCITEC.

Drying of Herbs, Spices, and Medicinal Plants

Edited by
Ching Lik Hii
Shivanand S. Shirkole

CRC Press
Taylor & Francis Group
Boca Raton London New York

CRC Press is an imprint of the
Taylor & Francis Group, an **Informa** business

Designed cover image: Shutterstock

First edition published 2024
by CRC Press
6000 Broken Sound Parkway NW, Suite 300, Boca Raton, FL 33487-2742

and by CRC Press
4 Park Square, Milton Park, Abingdon, Oxon, OX14 4RN

CRC Press is an imprint of Taylor & Francis Group, LLC

ISBN: 978-1-032-21616-4 (hbk)
ISBN: 978-1-032-21618-8 (pbk)
ISBN: 978-1-003-26925-0 (ebk)

DOI: 10.1201/9781003269250

Typeset in Times
by codeMantra

Contents

Series Preface

It is well known that the unit operation of drying is a highly energy-intensive operation encountered in diverse industrial sectors ranging from agricultural processing, ceramics, chemicals, minerals processing, pulp and paper, pharmaceuticals, coal polymer, food, forest products industries, as well as waste management. Drying also determines the quality of the final dried products. The need to make drying technologies sustainable and cost-effective via the application of modern scientific techniques is the goal of academic and industrial R&D activities around the world.

Drying is a truly multi- and interdisciplinary area. Over the last four decades, the scientific and technical literature on drying has seen an exponential growth. The continuously rising interest in this field is also evident from the success of numerous international conferences devoted to drying science and technology.

The establishment of this new series of books titled "Advances in Drying Science and Technology" is designed to provide authoritative and critical reviews and monographs focusing on current developments as well as future needs. It is expected that books in this series will be valuable to academic researchers as well as industry personnel involved in any aspect of drying and dewatering.

This series will also encompass themes and topics closely associated with drying operations, e.g., mechanical dewatering, energy savings in drying, environmental aspects, life cycle analysis, technoeconomics of drying, electrotechnologies, control and safety aspects, and so on.

Series Editor

Arun S. Mujumdar, PhD, is an internationally acclaimed expert in drying science and technologies. He is the Founding Chair in 1978 of the International Drying Symposium (IDS) series and Editor-in-Chief of *Drying Technology: An International Journal* since 1988. The fourth enhanced edition of his *Handbook of Industrial Drying* published by CRC Press has just appeared. He is the recipient of numerous international awards, including honorary doctorates from Lodz Technical University, Poland, the University of Lyon, France, and Western University, Ontario, Canada.

Please visit www.arunmujumdar.com for further details.

Preface

Herbs, spices, and medicinal plants are highly sought-after agricultural commodities in global market due to their unique flavor, aroma, and functional values (medicinal/therapeutic effects). The application of these plant products could be traced back to ancient civilizations such as Mesopotamia, Babylon, Egypt, China, India, Arabia, Greece, and Rome. Drying is a key unit operation for the purpose of preservation and to retain the key natural attributes/active compounds in these dried plant products. This book, *Drying of Herbs, Spices, and Medicinal Plants,* aims to produce an insight on the drying characteristics, dryer selection, physiochemical analyses, quality improvement, product development, storage, shelf life, safety, and economics evaluation. This book consists of ten chapters, and its specific features include:

- Comprehensive analysis on drying technologies and selection of dryers for herbs, spices, and medicinal plants.
- Impact of dryer's operating parameters on the dried product quality.
- Instrumentation and analysis of bioactive compounds in dried plant products.
- Storage, product development, and dried product quality.
- Techno-economic analysis and future development.

No prior knowledge of processing of herbs, spices, and medicinal plants is required, as this book is written for both general readers and researchers. Overall, this book is designed as a reference source for process plant engineers as well as research scientists who work in these related areas. This book gathers internationally recognized experts from the industry and academia, and they have provided the latest developments in drying technologies for herbs, spices, and medicinal plants, including updates from their own recent works.

We are grateful to the authors as well as the reviewers who have contributed their valuable time as well as their experience and expertise in the production of this book.

Ching Lik Hii
Shivanand S. Shirkole

Editors

Ching Lik Hii, PhD, is Associate Professor of Chemical Engineering and Co-Director of Future Food Malaysia Research Center at the University of Nottingham, Malaysia Campus. Before joining the university, he was Senior Research Officer (cocoa and chocolate processing) at the Chemistry and Technology Division, Malaysia Cocoa Board. Currently, he is also serving as Chair of the Food and Drink Special Interest Group (Malaysia Chapter), Institution of Chemical Engineers UK. Dr Hii's key research areas include cocoa processing, drying technology, mathematical modeling, and functional food processing. He is an editorial board member of *Malaysian Cocoa Journal*, *Journal of Applied Food Technology*, and *American Journal of Food Science and Technology*. Dr Hii is also a Professional Engineer (PEng) registered with the Board of Engineers Malaysia and a Chartered Engineer (CEng) registered with the Engineering Council UK.

Shivanand S. Shirkole, PhD, is Assistant Professor in the Department of Food Engineering and Technology, Institute of Chemical Technology Mumbai, ICT-IOC Odisha Campus, Bhubaneswar, India. He earned a PhD in 2020 at the Department of Food Process Engineering, National Institute of Technology, Rourkela, India. He has more than 4 years of industrial experience as a Plant Engineer at a Hyderabad-based multi-crop seed conditioning plant. His broad areas of research are low-moisture food safety and the thermal processing of food. Presently, he is working on sustainable technologies for food and agriproducts. He is an Executive Committee Member of the Association of Food Scientists and Technologists (India), NIT Rourkela Chapter, and working as an Associate Editor for the journal *Drying Technology* (Taylor & Francis).

Contributors

V. Archana
Department of Crop Management
Vanavarayar Institute of Agriculture
Coimbatore, India

Kosar Mohammadi Balili
Department of Food Science and
    Technology
Faculty of Agricultural Sciences
University of Guilan
Rasht, Iran

Nikita S. Bhatkar
Department of Food Engineering and
    Technology
Institute of Chemical Technology
    Mumbai
ICT-IOC Odisha Campus
Bhubaneswar, India

Eric Wei Chiang Chan
Department of Food Science and
    Nutrition
Faculty of Applied Sciences
UCSI University
Kuala Lumpur, Malaysia

Chien Hwa Chong
Department of Chemical and
    Environmental Engineering
Faculty of Science and
    Engineering
University of Nottingham Malaysia
Selangor Darul Ehsan, Malaysia

Lavanya Devaraj
Department of Food Technology
School of Agriculture and Food
    Technology
Vignan's Foundation for Science,
    Technology and Research
Guntur, India

Harshavardhan Dhulipalla
Department of Food Technology
School of Agriculture and Food
    Technology
Vignan's Foundation for Science,
    Technology and Research
Guntur, India

Nesa Dibagar
Institute of Agricultural Engineering
Faculty of Life Sciences and Technology
Wroclaw University of Environmental
    and Life Sciences
Wrocław, Poland

Adam Figiel
Institute of Agricultural Engineering
Wroclaw University of Environmental
    and Life Sciences
Wrocław, Poland

Ching Lik Hii
Department of Chemical and
    Environmental Engineering
Faculty of Science and Engineering
University of Nottingham Malaysia
Selangor Darul Ehsan, Malaysia

Fereshteh Jamalzade
Department of Food Science and
    Technology
Faculty of Agricultural Sciences
University of Guilan
Rasht, Iran

Hari Kavya Kommineni
Department of Food Technology
School of Agriculture and Food
    Technology
Vignan's Foundation for Science,
    Technology and Research
Guntur, India

Krzysztof Lech
Institute of Agricultural Engineering
Wrocław University of Environmental
    and Life Sciences
Wroclaw, Poland

Jacek Łyczko
Faculty of Biotechnology and Food
    Science
Wrocław University of Environmental
    and Life Sciences
Wrocław, Poland

Narjes Malekjani
Department of Food Science and
    Technology
Faculty of Agricultural Sciences
University of Guilan
Rasht, Iran

Klaudia Masztalerz
Institute of Agricultural Engineering
Wrocław University of Environmental
    and Life Sciences
Wroclaw, Poland

Rhonalyn Maulion
Department of Chemical
    Engineering
College of Engineering Architecture
    and Fine Arts
Batangas State University
Batangas City, Philippines

Kar Yong Pin
Department of Innovation and
    Commercialization
Forest Research Institute Malaysia
Selangor, Malaysia

Viplav Hari Pise
Department of Chemical
    Engineering
Institute of Chemical Technology
    Mumbai
Mumbai, India

Shivanand S. Shirkole
Department of Food Engineering and
    Technology
Institute of Chemical Technology
    Mumbai
ICT-IOC Odisha Campus
Bhubaneswar, India

Irshaan Syed
Department of Chemical Engineering
Vignan's Foundation for Science,
    Technology and Research (Deemed
    to be University)
Guntur, India

Fatemeh Poureshmanan Talemy
Department of Food Science and
    Technology
Faculty of Agricultural Sciences
University of Guilan
Rasht, Iran

Choon Hui Tan
Department of Food Science and
    Nutrition
UCSI University
Kuala Lumpur, Malaysia

Joash Ban Lee Tan
School of Science
Monash University Malaysia
Selangor Darul Ehsan, Malaysia

Bhaskar N. Thorat
Department of Chemical Engineering
Institute of Chemical Technology
    Mumbai
Mumbai, India

Vimal
Department of Food Engineering and
    Technology
Institute of Chemical Technology
    Mumbai
ICT-IOC Odisha Campus
Bhubaneswar, India

1 Overview of the Global Market for Dried Herbs, Spices, and Medicinal Plants

Ching Lik Hii
University of Nottingham Malaysia

Shivanand S. Shirkole
Institute of Chemical Technology Mumbai

CONTENTS

1.1 INTRODUCTION

Herbs and spices have been used since time immemorial as culinary ingredients as well as therapeutic applications. Various parts of the plants can be used for such applications, namely the arils, barks, berries, buds, bulbs, pistil, kernel, leaf, rhizome, latex, roots, and seeds. According to the International Organization for Standardization (ISO), spices and condiments are defined as vegetable products or mixtures thereof, free from extraneous matter, used for flavoring, seasoning, and imparting aroma to foods (ISO 676:1995). Generally, herbs and spices are differentiated based on the parts of the plants. Herbs are defined as the dried leaves of aromatic plants that are used in food for flavor enhancement, and spices are defined as the dried parts of aromatic plants excluding the leaves which covers practically all the plant parts (Peter and Shylaja, 2012). A complete list of the 109 herb and spice plant species can be viewed from the ISO 676:1995 Spices and condiments—Botanical nomenclature published by the International Organization for Standardization (https://www.iso.org). Figure 1.1 shows a general taxonomic classification of herbs and spices.

DOI: 10.1201/9781003269250-1

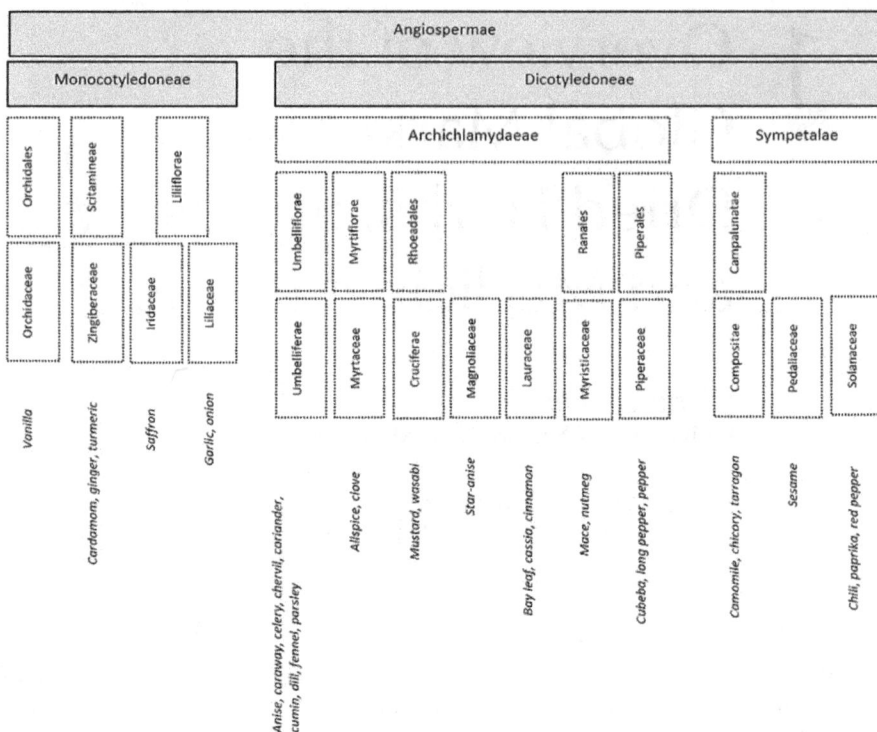

FIGURE 1.1   Herbs and spices from various plant parts. (Source: Peter and Shylaja, 2012.)

Historically, herbs and spices have been used not only in cooking but also for therapeutic purposes to treat body discomforts and for external applications. Ancient record on utilization of herbs and spices had also been documented in Mesopotamia, Babylon, China, India, Arabia, Greece, and Rome (Anon, 2022b). Ancient Egyptian medical literature *Ebers Papyrus* (1500 BC) cited medical treatments that consisted of caraway, coriander, fennel, garlic, mint, onion, peppermint, poppy, and onion. Typically, herbs and spices are known for several medicinal properties such as antimicrobial, anti-inflammatory, anticarcinogenic, antipyretic, antihelmintic, antispasmodic, anti-epileptic, antinauseant, aphrodisiac, carminative, analgesic, antiseptic, laxative, and so on. Table 1.1 shows mapping of the medicinal properties of various herbs and spices.

1.2   PRODUCTION AND CONSUMPTION

In 2021, world's production of spices and aromatic plants was 3,152,493.3 tons with 1,408,950 ha of land under cultivation. There is a marked increase in the trade of spices, herbs, and medicinal plants with an increase of 7.04% from 2019 to 2020. In 2020, total import and export of the spices and herbs were 6,299,556 and 6,580,052 tons, respectively, in the global markets (https://www.fao.org/faostat). Similarly, the

TABLE 1.1

Mapping of Medicinal Properties of Herbs and Spices

Herbs and Spices	Anticancer	Antihypertensive	Antidiabetic	Anti-Inflammatory	Anti-Obesity	Hepatoprotective	Digestive Stimulation	Hypolipidemic	Influence on Alzheimer's Disease
Black pepper (*Piper nigrum*)		x							
Celery (*Apium graveolens*)		x	x						
Cinnamon (*Cinnamomum zeylanicum*)			x		x	x			
Cumin (*Cuminum cyminum*)		x							
Fennel (*Foeniculum vulgare*)			x			x			
Ginger (*Zingiber officinale*)	x		x	x				x	
Rosemary (*Rosmarinus officinalis*)	x		x	x	x			x	x
Saffron (*Crocus sativus*)						x		x	
Turmeric (*Curcuma longa*)	x		x	x	x	x			x
Bay leaf (*Laurus nobilis*)			x						
Black cumin (*Nigella sativa*)	x				x				
Chilly (*Capsicum annuum* L.)									
Cloves (*Eugenia aromaticum*)			x					x	
Garlic (*Allium sativum*)	x							x	
Onion (*Allium cepa*)	x		x						
Sage (*Salvia officinalis*)									x
Others—mix of spices							x		

Source: Rodríguez-Pérez and Aznar (2020).

export of medicinal plant has grown by 33.2% in 2020–2021. The use of these commodities in the treatment of lethal diseases such as Covid 19 has also widen the
demand and scope of these commodities in the market (Nath and Debnath, 2022;
Singh et al., 2021). Producing herbs and spices for high-quality markets such as in
the Europe, USA, and United Arab Emirates generate opportunities (higher price
margins) as well as constraints (greater requirements on quality). In order to produce
medicinal plant products with good recovery of the phytochemicals, both harvesting
time and drying are of concern when it comes to the quality of these commodities
(Pandey and Savita, 2017). Herbs, spices, and medicinal plants are prone to high
microbiological load due to exposure to air, humidity, soil, dust, poor handling, and
storage practices. The most important microorganisms responsible for contamination
are *Salmonella spp.*, *Aspergillus flavus* producing aflatoxin, and *Aspergillus ochra-
ceus* producing ochratoxin (FDA, 2022, from https://www.fda.gov/food/natural-tox-
ins-food/mycotoxins). In 2013, United States Food and Drug Administration released
a three-year study report, which indicates that nearly 7% of imported spices from
India, Mexico, Thailand, and Vietnam were contaminated with *Salmonella*. During
this period, 749 shipments of spices were denied entry into the United States because
of *Salmonella* contamination, while 238 other shipments were denied because of the
presence of insects, excrement, hair, or other contaminants.

In general, most of the widely used herbs and spices are produced from the tropical regions (e.g., pepper, capsicums, nutmeg, cardamom, pimento, vanilla, cloves,
ginger, cinnamon, and turmeric), but some are also produced from non-tropical
regions (e.g., coriander, cumin, mustard, sesame seeds, sage, oregano, thyme, bay,
and mints). However, there are about 40–50 spices which are demanded highly for
its economic value and culinary importance. Consumption of herbs and spices varies according to locations with estimated consumption at 0.5 g per day in Europe,
1.8 g in parts of Africa, 1.3–1.9 g in Australia, and New Zealand, while in India,
South Africa, and Latin America recorded in average of 4.4 g per day. In terms of
consumption per capita, the highest consumption was recorded in Nepal (14 kg per
person), followed by Thailand (6.25 kg per person), Vietnam (4 kg per person), and
Turkey (3.71 kg per person). The world average was estimated at 1.69 kg per person
(Anon, 2020).

India, well known as home of spices, remains as the world's largest producer,
consumer, and exporter of spices and accounts for half of the global trading. During
2021–2022, the export of spices/spice products from India was about 1,531 K tons
valued at USD 4,102.29 million (Spices Board India, 2023). Total cultivation area
was estimated at 45.28 million ha and constituted about 9.6% of the total agriculture export earnings. China and Indonesia ranked second and third, respectively,
with production figures at 1.2 M tons and 659 K tons. India leads in the list as major
exporter, and other equally significant exporters are China, Vietnam, Indonesia,
Netherlands, Singapore, Germany, Brazil, Sri Lanka, and Guatemala. Major importing countries of spices are USA, Germany, Netherlands, India, Japan, UAE, UK,
Singapore, Spain, and France. In 2020, and based on continents, Asia was the largest
importer of spices and herbs (47%), followed by Europe (26%), North America (17%),
Africa (4.5%), Latin America, the Caribbean (3%), and Australia and Oceania (2.5%)
(Anon, 2022c).

1.3   INDUSTRIAL APPLICATIONS

Conventionally, herbs and spices are mainly used as natural flavoring ingredient in cooking. It can be used solely on its own or it can be made into seasoning or spice mixes. In the food and beverage industry, it is also used as natural colorant (e.g., yellowish color from saffron, red color from paprika, and green color from herbs) to replace synthetic colorant which could pose a health risk (e.g., carcinogenic). It is also used as antioxidants in the form of ground mixture, extract, emulsion, or encapsulated products. Some of the herbs and spices that were reported to contain high antioxidant activities are such as clove, ginger, cinnamon, turmeric, allspice, black pepper and cumin, oregano, sage, peppermint, thyme, rosemary, dill, and marjoram (Kurian, 2012). Table 1.2 shows antioxidant compounds that could be obtained from various herbs and spices.

Herbs and spices are also important ingredients for food preservation purposes due to its antimicrobial properties (e.g., seed saponins in fenugreek, carvacrol in oregano, and methyl chavicol in basil) that could inhibit the growth of food borne pathogens. Such properties have been discovered with high activity in oregano, cloves, cinnamon, garlic, coriander, rosemary, parsley, lemongrass, sage, and vanillin (Tajkarimi et al., 2010). Besides, herbs and spices also contain various health beneficial bioactive compounds such as vitamins, micro- and macroelements, glycosides, alkaloids, tannins, essential oils, flavonoids, antracompounds, phenols, coumarin, organic acids, and saponins.

Nowadays, many drugs and health products are formulated from herbs and spices that are known to possess medicinal properties. It is used not only in modern medicine practice but also in traditional and complementary medicine. It was reported that at least 80% of world population are relying on them as some form of primary health care (Ekor, 2014). The global herbal medicine market was estimated at USD 185.86 billion and is expected to grow to USD 430.05 billion in 2028. Pharmaceutical and nutraceutical sectors are the leading industries in this market followed by the food and beverage industries and personal care and beauty products industries (Fortune, 2022).

The World Health Organization (WHO) endorses utilization of herbal drugs in national healthcare programs as they are more affordable and can be accessed easily by the wider public especially the lower-income group. However, it should be noted that not all products can be claimed as prescription drugs and over-the-counter medicines unless they are tested through clinical trials and registered with the health authorities. Products other than this group are often known as health or herbal supplements, which are also classified as food. Some of the popular herbal supplements available in the market are black cohosh (used for menopausal conditions, painful menstruation, uterine spasms, and vaginitis), echinacea (used to strengthen the body's immune system), evening primrose (reducing symptoms of arthritis and premenstrual syndrome), feverfew (used for migraine headaches, as well as for menstrual cramps), garlic (used for cardiovascular conditions, including high cholesterol and triglyceride levels), gingko biloba (used for aging, including poor circulation and memory loss), ginseng (general tonic to increase overall body tone), goldenseal (healing properties and antiseptic, or germ-stopping qualities including for colds and flu), green tea (combat fatigue, prevent arteriosclerosis and certain cancers, lower cholesterol and weight loss), hawthorn (used for heart-related conditions and in the

TABLE 1.2

Antioxidant Compounds from Various Herbs and Spices

Herbs and Spices	Antioxidant Compounds
Bay leaf	Ascorbic acid, beta-carotene, tocopherols, eugenol, methyl eugenol, eudesmol, kaempferol, kaempferol-3-rhamnopyranoside, kaempferol-3,7-dirhamnopyranoside, 8-cineole, α-terpinyl acetate, terpinen-4-ol, catechin, cinnamtannin B1
Black pepper	Kaempferol, rhamnetin, quercetin, ascorbic acid, beta-carotene, ubiquinone, camphene, carvacrol, eugenol, gamma-terpinene, methyl eugenol, piperine
Chili pepper	Capsaicin, capsaicinol
Coriander	Beta-carotene, beta-sitosterol, caffeic acid, camphene, gamma-terpinene, isoquercitrin, myrcene, myristicin, p-hydroxy-benzoic acid, protocatechuic acid, quercetin, rhamnetin, rutin, scopoletin, tannin, terpinen-4-ol, trans-anethole, vanillic acid
Cumin	Cuminal, γ-terpinene, pinocarveol, linalool, 1-methyl-2-(1-methylethyl)benzene, carotol, apigenin, luteolin, cuminaldehyde, cuminic alcohol, p-cymene, β-pinene
Clove	Phenolic acids (gallic acid), flavonol glucosides, phenolic volatile oils (eugenol, acetyl eugenol, isoeugenol), tannins
Garlic	Caffeic, vanillic, p-hydroxybenzoic, and p-coumaric acids, allicin
Ginger	Gingerol, shogaol, zingerone, ascorbic acid, beta-carotene, caffeic acid, camphene, gamma-terpinene, p-coumaric-acid, terpinen-4-ol
Marjoram	Beta-carotene, beta-sitosterol, caffeic acid, carvacrol, eugenol, hydroquinone, linalyl-acetate plant, myrcene, rosmarinic acid, terpinen-4-ol, beta-carotene, caffeic acid, tannin, myrcene, phenol, trans-anethole, ursolic acid, oleanolic acid
Mustard	Carotenes, glucosinolates
Onion	Quercetin, kaempferol, cyanidin glucosides, peonidin glucosides, taxifolin, allicin
Oregano	Rosmarinic acid, caffeic acid, protocatechuic acid, 2-caffeoyloxy-3-[2-(4-hydroxybenzyl)-4,5-dihydroxy] phenylpropionic acid, flavonoids—apigen, eriodictyol, dihydroquercetin, dihydrokaempferol, carvacrol, thymol, camphene, gamma-terpinene, terpinen-4-ol, myrcene, linalyl-acetate
Peppermint	Ascorbic acid, beta-carotene, narirutin, eriodictyol, eriodictyol 7-O-β-glucoside, eriocitrin, hesperidin, isorhoifolin, luteolin 7-O-β-glucoside, luteolin 7-O-rutinoside, diosmin, rosmarinic acid, caffeic acid, piperitoside, menthoside, lithospermic acid
Rosemary	Carnosol, carnosic acid, rosmanol, rosmadial, diterpenes (epirosmanol, isorosmanol, and rosmaridiphenol), rosmariquinone, rosmarinic acid
Sage	Carnosol, carnosic acid, rosmanol, rosmadial, methyl and ethyl esters of carnosol, rosmarinic acid, ascorbic acid, beta-carotene, beta-sitosterol, camphene, gamma-terpinene, hispidulin, labiatic acid, oleanolic acid, terpinen-4-ol, ursolic acid, selenium, salvigenin, nevadensin, apigenin, cirsileol, cirsimaritin
Thyme	Thymol, carvacrol, p-cymene-2,3-diol, phenolic acids (gallic acid, caffeic acid, rosmarinic acid), phenolic diterpenes, flavonoids, ascorbic acid, beta-carotene, isochlorogenic acid, labiatic acid, p-coumaric acid, rosmarinic acid
Turmeric	Curcumins, 4-hydroxycinnamoyl methane, ascorbic acid, carotenes, caffeic acid, p-coumaric acid

Source: Leja and Czaczyk (2016).

treatment of angina, atherosclerosis, heart failure, and high blood pressure), saw palmetto (used for enlarged prostate), and St. John's wort (used in the treatment of mental disorders and depression) (Anon, 2022a). Consumers are advised to purchase only government authorities-approved herbal supplements in order to avoid adverse effect due to toxicity and contaminants issues caused by the products on user's health.

1.4   HARVESTING PRACTICES

In 2000, the demand for the medicinal and herbal plants surged drastically, and the demand was expected to grow, so since then many countries adopted the principles of good agricultural practices. In 2002, Herbal Medicinal Products of the European Medicines Agency put forth the guidelines on Good Agricultural and Collection Practices for these commodities. China adopts these practices in 2002. Similarly, WHO gave guidelines for the Good Agricultural and Collection Practices for Medicinal Plants in 2004. Furthermore, American Herbal Products Association laid down the Good Agricultural and Collection Practices and Good Manufacturing Practices for the Botanical Materials in 2017 (Zhang et al., 2021). These practices or guidelines mainly focus on the better handling of these commodities during pre-harvest period. The lower recovery of phytochemicals, determination of correct harvesting time and harvesting age, identification of the correct harvest part, and determination of the correct harvesting method are some of the crucial steps during the pre-harvesting period, which determines the quality, efficacy, safety, and other important aspects.

Similar to the medicinal plants, the harvesting of herbs and spices is totally dependent on the part of the plants of interest (Elsa Sánchez, 2007). During harvesting, care should be taken to avoid contaminations such as from the soil and fallen deteriorated commodities from the plant (IOSTA, 2008). The European Union provides certain guidelines for the specific maturity index of different plant parts to be harvested, and Table 1.3 shows the maturity time for the different plant parts.

TABLE 1.3
Maturity Time for the Different Plant Parts

Plant part	Maturity time
Leaves and aerial parts	Shortly before flowering.
Flowers	At full bloom, on sunny days and in the morning, when the dew has dried off.
Fruit and seeds	At full maturity, however, the seeds of some species tend to drop off or shatter.
Roots and rhizomes	At the beginning of senescence of the crop when the aerial parts are dead. Usually in late autumn or in early spring before sprouting when there is no frost and on dry days.

Source: Boor and Lefebvre (2022).

TABLE 1.4
Estimated Post-Harvest Losses of Red Pepper at Producer's Level

Causes	Losses (% of Total Production)
Due to high moisture	15–25
Due to spoilage in field	1–10
During farm to assembling	5–10
During assembling to distribution	2–5

Source: Ministry of Agriculture (2009).

The lower productivity of the herbs and spices is an alarming situation throughout the world. The major producers such as China and India are currently facing this issue, which will widen the gap between the demand and supply in the international market. The poor soil fertility, use of non-resistant variety, and higher use of chemical fertilizer are some of the factors responsible for the lower productivity (ICAR, 2013). The harvest commodities should be processed as early as possible to avoid losses and wastages these expensive commodities.

There are various post-harvest practices such as drying, sterilization, transportation, and storage that play a vital role in the production of quality spices and herbs and its by-products. For medicinal plants, the collection, processing, visual inspections, storage in ambient temperature after drying, packaging, transportation, and final storage are generally practiced. (Kurian and Sankar, 2007). However, a wide amount of wastage is associated with the post-harvest handling of these commodities; for instance, the estimated post-harvest spoilage of a red pepper at different post-harvest stages could be as high as 25% (Table 1.4).

1.5 CONCLUDING REMARKS

The consumption for dried herbs, spices, and medicinal plant products has increased considerably due to great demands from the food industries and general consumers, especially at this stage when markets start to recover from the COVID-19 pandemic. The harvesting and handling practices of these commodities are very crucial for reasons such as food safety, shelf life, and product quality. The supply and demand of these products are highly dependent on the regulatory bodies, which lay the guidelines and regulations in the international market. High attention should be paid to avoid contamination of these products with foreign matters, microbial load, and unwanted materials, and it is mandatory to meet the current international standards and criteria for these commodities.

REFERENCES

Anon. (2020). *The global spice market lacks to regain its former momentum.* Retrieved from h ttps://www.globaltrademag.com/the-global-spice-market-lacks-to-regain-its-former-momentum/

Anon. (2022a). *Herbal medicine.* Retrieved from https://www.hopkinsmedicine.org/health/wellness-and-prevention/herbal-medicine

Anon. (2022b). *History of spices*. Retrieved from https://www.mccormickscienceinstitute. com/resources/

Anon. (2022c). *What is the demand for spices and herbs on the European market?* Retrieved from https://www.cbi.eu/market-information/spices-herbs/what-demand

Boor, B., & Lefebvre, N. (2022). *Harvest and post-harvest handling of herbs*. Retrieved from https://orgprints.org/44077/2/1233-harvest-post-harvest-herbs.pdf

Ekor, M. (2014). The growing use of herbal medicines: issues relating to adverse reactions and challenges in monitoring safety. *Frontiers in Pharmacology*, 4, Article 177.

Elsa Sánchez, K. K. (2007). *Harvesting and preserving herbs and spices for use in cooking*. Retrieved from https://extension.psu.edu/harvesting-and-preserving-herbs-and-spices-for-use-in-cooking

Fortune. (2022). *Spices and seasonings market size, share and Covid-19 impact analysis*. Retrieved from https://www.fortunebusinessinsights.com/industry-reports/spices-and-seasonings-market-101694

ICAR. (2013). *Vision 2050*. Retrieved from http://spices.res.in/sites/default/files/Vision-IISR-2050.pdf

IOSTA. (2008). *General guidelines for good agricultural practices spices*. Retrieved from http://www.greenfoodec.eu/documents/general_guidelines_for_good_agric_prac-tices_spices.pdf

Kurian, A. (2012). Health benefits of herbs and spices. In K. V. Peter (Ed.). *Handbook of Herbs and Spices* (2nd ed., pp. 72–88). Sawston, UK: Woodhead Publishing.

Kurian, A., & Sankar, M. A. (2007). *Medicinal Plants* (Vol. 2). New Delhi, India: New India Publishing.

Leja, K. S., & Czaczyk, K. (2016). The industrial potential of herbs and spices - a mini review. *Acta Scientiarum Polonorum Technologia Alimentaria*, 15(4), 353–365.

Ministry of Agriculture. (2009). *Postharvest profile of chilli*. Retrieved from http://agmarknet. gov.in/Others/preface-chhilli.pdf

Nath, M., & Debnath, P. (2022). Therapeutic role of traditionally used Indian medicinal plants and spices in combating COVID-19 pandemic situation. *Journal of Biomolecular Structure Dynamics*, 1–20. doi:10.1080/07391102.2022.2093793.

Pandey, A., & Savita, R. (2017). Harvesting and post-harvest processing of medicinal plants: problems and prospects. *The Pharma Innovation Journal*, 6(12), 229–235.

Peter, K. V., & Shylaja, M. R. (2012). Introduction to herbs and spices: definitions, trade and applications. In K. V. Peter (Ed.). *Handbook of Herbs and Spices* (2nd ed., pp. 1–24). Sawston, UK: Woodhead Publishing.

Rodríguez-Pérez, C., & Aznar Roca, R. (2020). Medicinal properties of herbs and spices: past, present, and future. In M. B. Hossain et al. (Eds.). *Herbs, Spices and Medicinal Plants: Processing, Health Benefits and Safety* (1st ed., pp. 207–249). New York: John Wiley & Sons Ltd.

Singh, N.A., Kumar, P., Jyoti Kumar, N. (2021). Spices and herbs: Potential antiviral preventives and immunity boosters during COVID-19. *Phytotherapy Research,* 35(5), 2745–2757. doi: 10.1002/ptr.7019. Epub 2021 Jan 29. PMID: 33511704; PMCID: PMC8013177.

Spices Board India (2023). *Trace information and statistics*. Retrieved from http://www.indi-anspices.com/box2info.html

Tajkarimi, M. M., Ibrahim, S. A., & Cliver, D. O. (2010). Antimicrobial herb and spice compounds in food. *Food Control*, 21, 1199–1218.

Zhang, M., Wang, C., Zhang, R., Chen, Y., Zhang, C., Heidi, H., & Li, M. (2021). Comparison of the guidelines on good agricultural and collection practices in herbal medicine of the European Union, China, the WHO, and the United States of America. *Journal of Pharmacological Research*, 167, 105533.

2 Guideline and Selection of Dryers for Herbs, Spices, and Medicinal Plants

Choon Hui Tan
UCSI University

Ching Lik Hii
University of Nottingham Malaysia

CONTENTS

2.1 INTRODUCTION

Drying of food products plays an important role as an essential operation to enhance food security. The removal of moisture to safe level primarily retards moisture-mediated deteriorative biochemical reactions and prevents the growth/reproduction of microorganisms. Furthermore, drying can also reduce the product's weight, costs of transportation, and packaging, as well as extend product's shelf life during storage and preserve important nutrients (Mujumdar, 1995). Various types of herbs and spices are traded in markets worldwide, and majority are processed into semi- or fully dried forms. In the culinary sense, dried herbs and spices are generally used to improve the aroma, taste, and functionality of processed food. Besides, herbs and spices can also be applied in medicine, pharmacology, and cosmetology owing to their medicinal properties, antioxidant activities and nutritional properties (Peter, 2012). For example, some herbs and spices have been reported able to exhibit anti-cancer, anti-microbial, anti-hypertensive, anti-diabetic, anti-inflammatory and anti-obesity properties. Traditional and complementary medicine practitioners usually

DOI: 10.1201/9781003269250-2

apply or prescribe treatment that involves these medicinal plant products. Such treatments are still very popular especially in China and India where the highest number of medicinal plants used had been recorded.

Herbs and spices are highly perishable foodstuffs as they have high moisture content and majority are also chill-sensitive (Pirbalouti et al., 2013). The high level of moisture content is conducive for microbial growth and will cause further spoilage. Therefore, drying is a viable option to extend the shelf life to safeguard the important nutrients and health beneficial ingredients in the dried products. Besides, herbs and spices including medicinal plants in dried forms could facilitate grinding of the products and improve process efficiency. Moreover, drying could assist in the release of volatile aromatics, which could be further broken down and recombined to form different kind of compounds that increase the complexity of spices (Sankhé, 2019).

The aim of this chapter is to provide a guideline for the selection of dryers that can be used in processing of herbs, spices, and medicinal plants. The first section discusses different conventional dryers commonly used and includes comparison with some of the advanced dryers being currently developed. Subsequent sections will focus on the selection guide for dryers based on heat sensitiveness of bioactive compounds and drying efficiencies. Future development and recommendation of dryers will also be presented.

2.2 DRYING TECHNOLOGIES

In general, most industrial dryers that are used in food-processing operations are applicable to herbs, spices, and medicinal plants. Description of various types of dryers and their operation has been reported in several literature studies (Sagar and Kumar, 2010; Hii et al., 2015; Zhang et al., 2017). Conventionally, most herbs and spices are dried using direct sunlight as most of the plants are grown in tropical regions where solar radiation is abundant and the number of sunshine hours is higher during daytime. For industrial operation, hot air dryers are used to cater for the bigger quantity and also to ensure better hygiene and more consistent product quality. The dryer can be operated either in batch mode or in continuous mode. In both cases (open sun and hot air drying), the fresh materials are usually put in perforated trays to facilitate drying. However, both open sun and hot air drying are often associated with several drawbacks, as described in Table 2.1.

As drying technology progresses, more advanced dryers have been developed with better efficiency and being able to retain higher concentration of bioactive ingredients and flavor/aroma. Such dryers include solar dryer, vacuum dryer, freeze dryer, heat pump dryer, microwave dryer, infrared dryer, and so on. However, the dryers that utilize only single heating/moisture removal mechanism might not be that efficient after all, both in terms of drying efficiency and product quality. For example, solar dryer discontinues drying at night due to the absence of sunlight and freeze dryer operates under vacuum condition, which requires long operating hours. Recently, hybrid drying has gained interests from both researchers and industrial practitioners as such technique is able to combine advantages from two or more heating/drying mechanisms, which results in a more synergistic effects particularly in enhancement

TABLE 2.1
Drawbacks of Open Sun and Hot Air Drying

Open Sun Drying	Hot Air Drying
• Drying duration depends highly on weather condition	• High temperature/fast drying causes severe case hardening
• Low throughput	• Unappealing appearance/high shrinkage
• Contamination from foreign materials/debris	• Slow/poor rehydrability
• Overheating under too intense sunlight	• Weak flavor/aroma
• Susceptible to insect infestation	• Hard texture
• Susceptible to moisture re-absorption/mold growth	• Low retention of nutrients/bioactive ingredients
• Unappealing appearance/color	• High energy consumption

of drying/energy efficiency and product quality. Physical field-based technologies (Figure 2.1a–d) are some of the popular combinations that utilize energy from infrared, microwave, ultrasound, and radio frequency (Hii et al., 2021).

Most herbs, spices, and medicinal plants contain heat-sensitive ingredients (e.g., aroma and flavor compounds, nutrients, and antioxidants) that need to be processed with care. Too harsh drying conditions could negatively affect the retention of these ingredients especially due to extended heating at high temperature, which subsequently leads to further degradation. Ultimately, consumers perceive product quality as the main deciding factor for purchasing the dried products or at least the products should meet the acceptable benchmarks set by the market.

Hence, the following selection criteria provide a guideline to selection of dryers based on drying/energy efficiency and heat sensitivity, which are equally important parameters in production of high-quality herb, spice, and medicinal plant products.

2.3   GUIDELINES AND SELECTION

2.3.1   ENERGY EFFICIENCY

Drying processes consume great amount of energy which is mainly used for moisture evaporation purposes as the latent heat of vaporization of water is very much higher (2,260 kJ/kg) than the sensible heat. Statistics showed that industrial drying constituted about 10%–20% of the national industrial energy consumption in the developed world (Lee et al., 2013). Poor dryer design also causes energy losses within the range of 10%–40% due to ineffective insulation, exhaust gases, and other inadequate design features. The efficiencies determined also vary among dryers used; for example, moderate to high efficiency was recorded in hot air dryer (35%–40%), vacuum (70%), and heat pump (95%) dryers but freeze dryer showed a much lower efficiency (10%) among all (Perera and Rahman, 1997). In general, drying efficiency can be determined from several energy performance indicators such as energy efficiency, specific moisture extraction rate (SMER), and specific energy consumption (SEC) as shown in Table 2.2.

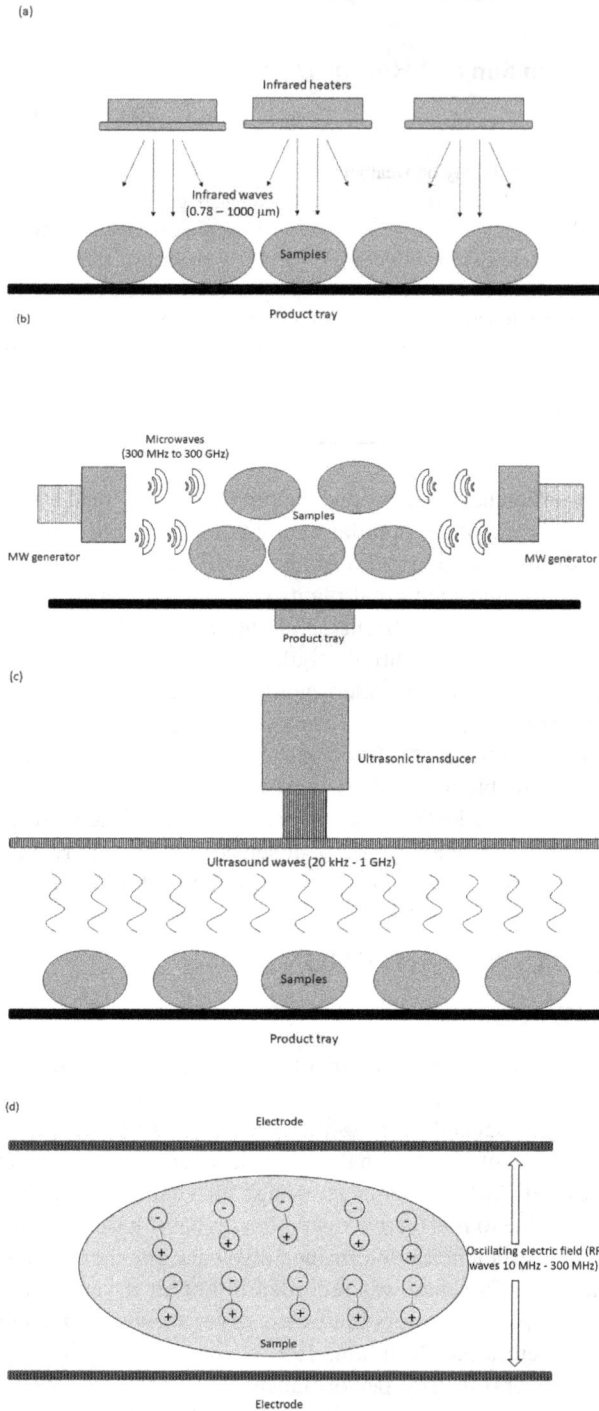

FIGURE 2.1 Examples of physical field-based drying (a) infrared, (b) microwave, (c) ultrasound, (d) radio frequency.

TABLE 2.2

Energy Performance Indicators for Dryers

Performance Indicator	Equation	Parameter
Energy efficiency (η)	$\eta = \dfrac{\varepsilon_r}{\varepsilon_s}$	ε_r = energy required (J) and ε_s = energy supplied (J)
	$\eta = \dfrac{\varepsilon_r + \varepsilon_h}{\varepsilon_s}$	ε_h = energy used for product heating (J)
Specific moisture extraction rate (SMER)	$\text{SMER} = \dfrac{M_w}{E}$	M_w = mass of water evaporated (kg) and E = energy usage (kWh)
Specific energy consumption (SEC)	$\text{SEC} = \dfrac{E}{M_i}$	M_i = initial mass of sample (kg) and E = energy usage (kWh)

Table 2.3 presents selected examples of drying/energy efficiencies reported for various dryers investigated by researchers. Current trends in food-processing technologies are gearing toward the applications of advanced drying using hybrid technologies such as combining microwave, infrared, vacuum or heat pump drying with conventional dryers (e.g., solar, hot air, and freeze dryers). It can be seen that upon hybridization, the dryers outperformed conventional dryers as shown by various energy performance indicators. The improved efficiencies are mainly attributed to several factors such as drying being more uniform/targeted at the product (e.g., microwave and infrared heating), energy saving scheme by incorporating thermal storage (e.g., solar drying), energy recovery mechanism (e.g., heat pump system), and optimized drying conditions (e.g., air flow, heating power, temperature, product size, and loading).

Most herbs, spices, and medicinal plant materials in their natural forms (e.g., leaves, bark, seed, and root) and also those processed into granular or powdered forms are suitable to be dried using hybrid dryers. Existing conventional dryers can also be hybridized either in combined mode or in sequential mode and require only moderate upgrade to the process control and product handling. Hence, it is worth to consider hybrid drying as compared to conventional/single drying strategy to achieve better efficiency that would help to improve the economics in processing of herbs, spices, and medicinal plants.

2.3.2 Heat Sensitivity

It is well established that depending on the type of raw material, the selection of an appropriate dryer is crucial as excessive heating at high temperature could impact substantially the physical, chemical, biological, and nutritional characteristics of the dried products. The heat sensitiveness of different herbs, spices, and medicinal plants as affected by drying is presented in Table 2.4. Some examples of the heat-sensitive compounds in herbs and spices are such as chlorophyll, antioxidants, vitamins, and so on.

TABLE 2.3

Energy Efficiency of Herbs, Spices, and Medicinal Plants Drying

Technique	Sample	Energy Performance Indicator	Reference
Forced convection SL dryer	Valerian rhizomes (*Valeriana officinalis* L.)	SL dryer without SHSM and PCM • Maximum heat collected by the collector was 294.5 W • Maximum collector efficiency was 37.84% • Drying efficiency was 12.92% SL dryer with SHSM without PCM • Maximum heat collected by the collector was 628.9 W • Maximum collector efficiency was 78.02% • Drying efficiency was 21.58% SL dryer with SHSM and PCM • Maximum heat collected by the collector was 628.9 W • Maximum collector efficiency was 78.02% • Drying efficiency was 27.84%	Bhardwaj et al. (2021)
Mixed mode SL dryer	Black ginger	Case 1 (constant air mass flow rate of 0.062 kg/s throughout drying • Average efficiency of solar air collector no.1 was 75% and for no.2 was 50% • Overall average efficiency of both collectors was 60% • Average dryer efficiency was 6.4%, and SEC was 1.07 kWh/kg Case 2 (two successive mass flow rates of 0.062 kg/s during the initial period of drying and 0.018 kg/s during the falling rate period) • Average efficiency of solar air collector no.1 was 58.3% and for no.2 was 40.6% • Overall average efficiency of both collectors was 49.4% • Average dryer efficiency was 10.8%, and SEC was 0.56 kWh/kg	Ekka et al. (2020)
MW-HA tunnel dryer	Onion slices	Energy efficiency: • MW-HA tunnel drying (9.38%–78.8%) • HA tunnel drying (3.45%–5.07%)	Maftoonazad et al. (2020)
HP dryer	Jew's mellow (*Corchorus olitorius*)	SEC decreases with the following drying conditions: • Surface load (14–28 kg/m^2) from 4,992 to 3,759 kJ/kg water. • Temperature (45°C–55°C) from 5,430 to 3,695 kJ/kg water. • Velocity (1.2–2.7 m/s) from 3,759 to 2,695 kJ/kg water. • Herb size (large–small) from 6,469 to 4,309 kJ/kg water. • Stem (stem–stemless) from 4,029 to 3,695 kJ/kg water.	Fatouh et al. (2006)

(Continued)

TABLE 2.3 (*Continued*)

Energy Efficiency of Herbs, Spices, and Medicinal Plants Drying

Technique	Sample	Energy Performance Indicator	Reference
SL dryer with thermal storage	Black pepper	Case 1 (product received thermal energy from both solar air heaters and top transparent cover) • Overall dryer efficiency (42.5%) • SEC (2.2 kWh/kg) Case 2 (top portion was completely insulated and the product received heat only from the solar air heaters) • Overall dryer efficiency (53.1%) • SEC (3.2 kWh/kg) Energy supplied from thermal storage recorded at 2.3 kW (both case 1 and 2)	Lakshmi et al. (2021)
HA, MWV drying and HA-MWV dryers	Curry leaf	SEC: • Convective HA drying (174.78–676.27 kJ/g initial mass) • MWV drying (35.43–85.49 kJ/g initial mass) • HA followed by MWV drying (161.6 kJ/g initial mass)	Choo et al. (2020)
HP-photovoltaic-thermal solar dryer	Saffron	• The highest drying efficiency and SMER were recorded at 72% and 1.16 (air temperature = 60°C and air flow rate = 0.016 kg/s), respectively • Usage of heat pump reduced total energy consumption by 33%.	Mortezapour et al. (2012)
HA and MW dryers	Peppermint	HA dryer (50°C–70°C): • SEC (42.72–64.71 MJ/kg water) • Thermal efficiency (3.69%–5.47%) • Energy efficiency (3.55%–5.34%) MW dryer (200–800 W): • SEC (4.58–5.89 MJ/kg w) • Thermal efficiency (38.36%–49.33%) • Energy efficiency (38.39%–49.33%)	Torki-Harchegani et al. (2016)
HA, IR, and IR-HA air dryers	Mint	• A minimum energy consumption of 2.31 kWh was recorded utilizing hybrid dryer and lower than hot air and infrared dryer by 93.3% and 49.6%, respectively • SEC (hybrid drying) reduced according to temperature (30°C–50°C) and infrared intensity (1,000–4,000 kWh) due to greater moisture evaporation • SEC (hybrid drying) increased with air velocity increment (1.0–2.5 m/s) due to the additional cooling effect	Ye et al. (2021)

(*Continued*)

TABLE 2.3 (*Continued*)
Energy Efficiency of Herbs, Spices, and Medicinal Plants Drying

Technique	Sample	Energy Performance Indicator	Reference
FZ, CV, and MW dryers	Purple basil	FZ dryer (−48°C, 13.33 Pa) • SMER (0.001 kg/kWh) • MER (0.001 kg/h) • SEC (709.09 kWh/kg) CV dryer (40°C–55°C, 1.5 m/s) • SMER (0.008–0.012 kg/kWh) • MER (0.021–0.032 kg/h) • SEC (81.34–124.21 kWh/kg) MW dryer (350–600 W) • SMER (0.334–0.356 kg/kWh) • MER (0.188–0.332 kg/h) • SEC (2.81–2.99 kWh/kg)	Altay et al. (2019)
CV, IR, CV–IR, MW, MW–CV, MWV, VC, and SL (with/ without HP) dryers	Chamomile	CV dryer (40°C–60°C) • Drying efficiency (2.21%–9.33%) • Thermal efficiency (2.12%–6.87%) IR dryer (0.22–0.49 W/cm² and 0.5–1.5 m/s) • Drying efficiency (3.71%–13.97%) • Thermal efficiency (4.03%–12.96%) CV-IR dryer (0.22–0.49 W/cm² and 0.5–1.5 m/s) • 40°C—drying efficiency (3.51%–18.03%) and thermal efficiency (3.63%–16.35%) • 50°C—drying efficiency (3.15%–19.93%) and thermal efficiency (3.16%–17.83%) • 60°C—drying efficiency (3.19%–23.34%) and thermal efficiency (3.14%–20.57%) MW dryer (100–1,000 W) • Drying efficiency (48.51%–65.24%) • Thermal efficiency (35.83%–49.99%) MWV dryer (130–450 W and 25–750 mbar) • Drying efficiency (12.41%–50.43%) • Thermal efficiency (37.96%–78.21%) MW-CV dryer (100–400 W and 40°C–60°C) • Drying efficiency (14.05%–33.47%) • Thermal efficiency (13.37%–24.66%) VC dryer (25–750 mbar and 40°C–70°C) • Drying efficiency (1.65%–10.24%) • Thermal efficiency (2.91%–11.94%) SL drying (40°C–60°C and 0.5–1.5 m/s) • Without heat pump—drying efficiency (4.14%–17.11%) and thermal efficiency (4.41%–13.17%) • With heat pump—drying efficiency (6.54%–23.52%) and thermal efficiency (6.98%–18.11%)	Motevali et al. (2018)

CV, convective air; FZ, freeze drying; HP, heat pump; HA, hot air; IR, infrared; MW, microwave; MWV, microwave vacuum; SL = solar.

TABLE 2.4

Heat Sensitivities of Herbs, Spices, and Medicinal Plants

Sample	Technique	Outcomes (Heat Sensitiveness)	Reference
Coriander leaves	Low temperature low humidity (LTLH) drying ($T = 50°C$, $60°C$, $80°C$ and $100°C$)	• LTLH-dried coriander ($T = 50°C$, humidity $= 25\%$–30% for 270 minutes) showed higher retention of total chlorophyll (5.3 g/100g), carotenoids (27.3 mg/100g), ascorbic acid (110.5 mg/100g) and greenness (-6.73) compared to other drying conditions.	Khanum et al. (2013)
Kaffir lime leaves	Low relative humidity (LRH) air drying ($T = 50°C$, air velocity $= 1.5$ m/s and RH $= 21\%$) Far-infrared radiation (FIR) drying ($T = 50°C$, air velocity $= 1.5$ m/s, FIR intensity $= 5$ kW/m^2)	• Total phenolic content (39% vs 62%) total flavonoid content (43% vs 39%) and antioxidant power (28% vs 39%) were found increased comparing between LRH and FIR drying. • Slight decrease in $L*$ values was observed in both LRH- and FIR-dried samples • LRH air drying and FIR drying increased the concentrations of citronellal (83% and 64%, respectively) with reference to the concentration in fresh leaves (56%).	Raksakantong et al. (2012)
Pecah beling leaves (*Strobilanthes crispus*)	Microwave vacuum drying (MWV) at power $= 6$, 9 and 12 W/g Freeze drying (FD)	• The highest antioxidant activity (4.47 µM Trolox/100g), total volatiles and phytosterols were obtained with MWV at 9 W/g. • In MWV, total phenolic content considerably decreased from 904.05 to 830.37 mg/100g dw when microwave power increased to 12 W/g. • MWV and FD caused losses of total phenolic compound and retained 873.63 mg/100g dw and 904.05 mg/100g dw, respectively, as compared to the fresh samples at 1,222.33 mg/100g dw. • FD samples had the lightest (highest $L*$) and greenest (lowest $a*$) color profiles.	Chua et al. (2019)

(Continued)

TABLE 2.4 (*Continued*)
Heat Sensitivities of Herbs, Spices, and Medicinal Plants

Sample	Technique	Outcomes (Heat Sensitiveness)	Reference
Peppermint	Natural drying Convection drying ($T = 39°C$ and air flow rate = 1.5 m/s) Infrared drying ($T = 39°C$ and wavelength = 1.2–6.0 µm) Vacuum drying ($T = 30°C$ and pressure = 7×10^{-6} mPa) Microwave drying (Frequency = 2,450 MHz and power = 500 W) Sublimation drying (condenser temperature = $-85°C$ and vacuum = 2×10^{-6} mPa)	• MW samples retained the lowest essential oil content. • MW samples showed the biggest changes in greenness $a*$ value. • The positive value of greenness $a*$ in the cv. "Peppermint" showed that the MW samples had a reddish tint. • Chroma values of the MW-dried peppermint leaves did not reduce. • Large hue angle (h°) changes were observed in IR, convective, and MW-dried samples, mostly in cv. "Peppermint."	Rubinskiene et al. (2015)
Linden leaves	Infrared drying ($T = 50, 60$ and 70°C)	• $L*$ and $b*$ values of the dried leaves increased partially due to the degradation of chlorophyll. • The highest $L*$ (42.36) and $b*$ (30.56) were obtained at 60°C. • Drying temperature at 50°C is recommended in terms of total phenolic content and total flavonoid content retention (95.12 mg/g DW and 2.79 mg/g DW, respectively) as compared to the fresh samples.	Selvi (2020)
Dill leaves	Infrared drying (IR power = 1,790, 1,970 and 2,070 W/m²)	• The brightness ($L*$), greenness ($a*$) and yellowness ($b*$) values of dried samples were between 8.37 and 10.91, −13.57 and −11.57, 6.36 and 7.36, respectively, when infrared power intensity increased. • Changes of color parameters could be due to the chlorophyll degradation and resulted in lower chlorophyll retention.	Tezcan et al. (2020)

(Continued)

TABLE 2.4 (*Continued*)
Heat Sensitivities of Herbs, Spices, and Medicinal Plants

Sample	Technique	Outcomes (Heat Sensitiveness)	Reference
Ginger	Freeze drying ($T = -80°C$) Catalytic infrared drying ($T = 60°C$)	• Catalytic infrared drying retained the bioactive ingredients better in the dried products as the reduction rates were 2.04% and 3.69%, respectively, for total phenolic and total flavonoid compounds. • It showed lesser damage to the cortical tissue, with well-preserved cell structures. • Freeze drying showed a slight damage to the cell wall but good retention of intact starch granules in cells. • Freeze-dried samples had the highest free radical elimination rate with DPPH and ABTS free radical elimination rates at 40.14% and 40.39% higher than the fresh samples.	Ren et al. (2021)
Pepper	Microwave vacuum drying (MWV) at 800 W and 5 Pa Freeze drying (FD)	• MWV drying was found efficient in minimizing the loss of nutritional attributes. • FD was observed efficient in minimizing the loss of color, capsaicin and β-carotene. • FD-, MVD-, and sun-dried samples were found to cause chlorophyll degradation as the color parameters of fresh sample was the highest ($L^* = 49.24$, $a^* = 35.01$ and $b^* = 24.62$).	Maurya et al. (2018)
Tongkat Ali (*Eurycoma longifolia*)	Spray drying ($T = 100°C$, 180°C and 220°C)	• Spray-dried samples at 180°C and 220°C showed greater process yield (39.2% and 39%, respectively), lower moisture contents (8.81% and 5.90%, respectively), produced non-sticky particles. • It showed good powder size distribution (5.27 μm and 4.97 μm, respectively) of the Tongkat Ali extract as compared to spray drying at 100°C.	Harun et al. (2015)

(Continued)

TABLE 2.4 *(Continued)*
Heat Sensitivities of Herbs, Spices, and Medicinal Plants

Sample	Technique	Outcomes (Heat Sensitiveness)	Reference
Taegeuk ginseng (*Panax ginseng*)	Far-infrared drying (FIR) at 45°C, 55°C and 65°C and air velocity at 0.6 m/s	• The total color changes (ΔE) of hot air-dried sample were 11.91%–20.05% higher than FIR-dried samples. • FIR-dried samples showed higher retention of saponin (1.01 mg/g) and polyphenol contents (7.81 mg/100g) than hot air-dried samples. • FIR drying temperature should maintain below 55°C to obtain optimal conditions of dried Taegeuk ginseng with antioxidant level and polyphenols content of 110.34 mg (AEAC)/100 g and 164.85 mg/100g, respectively.	Ning and Han (2013)
Shiitake mushroom	Microwave vacuum drying (MWV) (power = 56, 143, 209 and 267 W; pressure = 18.66, 29.32, 39.99 and 50.65 kPa)	• MWV drying caused reduction of lightness (L^*) from 25% to 10% among all microwave power settings. • The minimum total color change (ΔE) of MWV sample was recorded at 6% using 267 W and 18.66 kPa.	Kantrong et al. (2014)
Stevia rebaudiana leaves	Microwave drying (power = 800 W and t = 4–8 minutes) Infrared drying (T = 60°C and t = 2–4 hours) Vacuum drying (T = 60°C and p = 15 kPa) Freeze drying (T = −18°C for 10 hours)	• Infrared-dried sample showed the lowest total chlorophyll at 0.402 g/100g DM, while freeze-dried sample showed the highest at 0.646 g/100g DM. • Freeze-dried (116.3 g/100g DM), vacuum-dried (122.7 g/100g DM), and microwave-dried (117.2 g/100g DM) samples had higher vitamin C value as compared to other drying treatments (sun, shade and infrared drying). • Freeze-dried sample was found to maintain the highest L^* and b^* values (50.21 and 24.5, respectively) as compared to other drying method.	Lemus-Mondaca et al. (2016)
Peperomia	Heat pump drying (T = 40°C, 45°C, 50°C and 55°C)	• Flavonoid content of dried samples decreased from 0.95 to 0.85 mg/mL as temperature increased from 40°C to 55°C. • Vitamin C content of dried samples decreased from 30.02 to 28.89 mg/100g as temperature increased from 40°C to 55°C.	Minh (2019)

(Continued)

TABLE 2.4 (*Continued*)
Heat Sensitivities of Herbs, Spices, and Medicinal Plants

Sample	Technique	Outcomes (Heat Sensitiveness)	Reference
Black pepper	Solar drying (Case 1: transparent top portion and act as mixed mode; Case 2: Insulated top portion and heating only from solar air heater; Case 3: sun drying)	• Total phenolic content for dried samples of case 1, 2, and 3 were 29.87, 30.11, and 18.35 mg GAE/g, respectively. • The antioxidant contents of samples dried in case 1 and case 3 were 45.69 µ mol TE/g and 23.25 µ mol TE/g, respectively.	Lakshmi et al. (2021)
Chili pod	Solar drying ($t = 57$ hours)	• Beta-carotene and vitamin C content of whole chili pods dried sample using desiccant in solar drying were the highest (20.38 and 23.99 mg/100g, respectively, as compared to natural convection solar drying. • Color attributes (L^*, a^* and b^*) of half chili pods dried sample without using desiccant in solar drying remained the highest ($L^* = 30.06$, $a^* = 20.26$ and $b^* = 8.49$).	Romauli et al. (2021)

Many herbs and spices contain chlorophyll, which could be degraded when exposed to high temperature environment. It has low stability during drying and degradation could result in the formation of pheophytins, which change the color from green to olive brown. Such changes are often reflected in the reduction of the green (a^*) and yellow (b^*) color attributes which in turn reduce the resultant chroma (Ch) attribute. However, the level of degradation is highly dependent on the species, and typically materials from the same family would show similar stability (Śledź and Witrowa-Rajchert, 2012).

Besides, polyphenols have been reported as heat-sensitive and prolonged heat treatment might cause irreversible chemical reaction to occur (Mejia-Meza et al., 2008). The reduction of total phenolic compounds might be attributed to the binding of polyphenols with other compounds or alteration of polyphenols' chemical structure (Martín-Cabrejas et al., 2009; Qu et al., 2010). Drying generally causes the reduction in antioxidant activity, which could be due to the loss of antioxidants or formation of compounds with pro-oxidant action (Sharma et al., 2015). As flavonoids are heat-sensitive, heating may cause breakdown of phytochemicals in herbs and spices, which affect cell wall integrity and cause migration of the flavonoid components. Oxygen, enzymes, and light might be the reason contributing to flavonoids degradation (Davey et al., 2000) but in some cases, the antioxidant activity increased which could be due to the alteration of the existing antioxidant structure and formation of novel antioxidant compounds in the dried herbs and spices (Gorinstein et al., 2008; Jiménez-Monreal et al., 2009).

In terms of phytonutrients, high temperature drying caused intense degradation of beta-carotene, which could be due to the reduction in the amount and bioactivity of phytonutrients caused by the enzymatic degradation process, and subsequently caused the change in the composition in phytonutrients (Muller-Harvey, 2001). Suvarnakuta et al. (2005) reported that degradation of carotenoid content was due to the activation of lipoxygenase and peroxidase. Similarly, the ascorbic acid (vitamin C) content decreased after heat treatment due to irreversible oxidative reaction and light energy that promoted destruction of ascorbic acid (Vega-Gálvez et al., 2009; Duncan and Chang, 2012). Vitamin C degraded when exposed to light energy by quenching singlet oxygen before it reacts with other compounds (Chou and Khan, 1983; Jung et al., 1995). Besides, heat treatment was known to promote the oxidation of ascorbic acid to dehydroascorbic and polymerized to other nutritional inactive components (Ayusuk et al., 2009).

2.4 RECOMMENDATION

Based on the comparison made between dryer types, drying/energy efficiencies and impact of drying on the heat-sensitive compounds in herbs, spices, and medical plants, it can be seen that the synergetic effect of hybrid drying could produce dried products with high retention of various health beneficial ingredients and with drying efficiencies better than conventional drying methods. Figure 2.2 shows recommended selections of hybrid dryers based on literature studies that are suitable for herbs, spices, and medicinal plants.

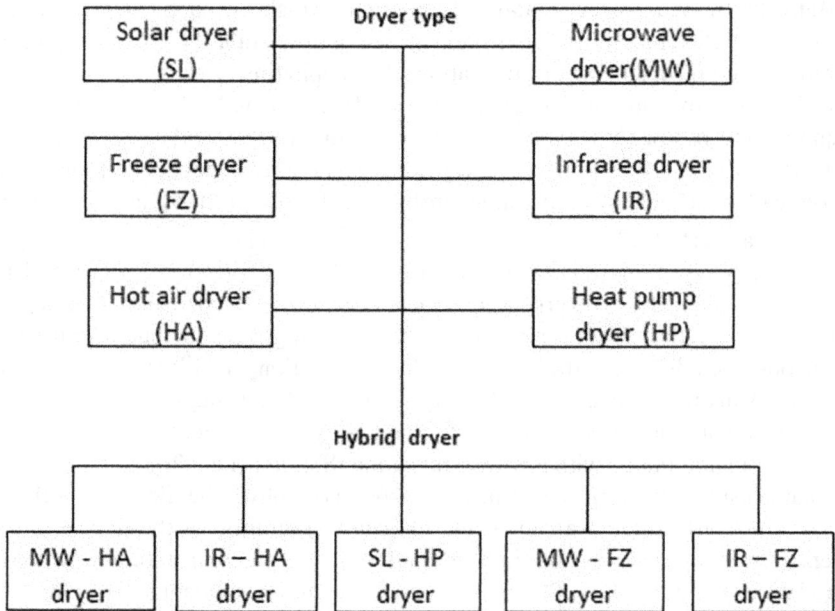

FIGURE 2.2 Recommended selections of hybrid dryers.

The benefits of hybrid drying have been reported in several literatures (Onwude et al., 2017; Li et al., 2019; Hii et al., 2021). It is also recommended that hybrid dryer is further improvised to maximize the combined heating mechanisms and efficiencies. For example, microwave vacuum dryer can be further improved by incorporating infrared heating. The coupled effect of microwave and infrared as well as the operation under vacuum can further promote moisture removal and improve drying efficiency. Likewise, heat pump-assisted solar dryer can also be combined with a fluidized bed system to facilitate product mixing. Fluidization is known able to improve heat and mass transfer during drying. As a result, this will further promote drying uniformity and produce better quality dried products.

2.5 CONCLUDING REMARKS

Drying is one of the important unit operations in processing of herb, spice, and medicinal plant products. Selection of suitable moisture removal mechanism as affected by the heating mechanisms is crucial both in safeguarding the heat sensitivity of the key compounds and also in maintaining efficient drying operation. Hybrid drying technologies have been proven able to achieve these requirements, and it is highly recommended that existing dryer be modified to incorporate additional heating mechanism and convert into hybrid dryer. The synergistic effect from hybrid dryers could contribute to better sustainability in drying operations over time.

REFERENCES

Altay, K., Hayaloglu, A.A., & Dirim, S.N. (2019). Determination of the drying kinetics and energy efficiency of purple basil (*Ocimum basilicum* L.) leaves using different drying methods. *Heat and Mass Transfer*, *55*, 2173–2184. https://doi.org/10.1007/s00231-019-02570-9

Ayusuk, S., Siripongvutikorn, S., Thummaratwasik, P., & Usawakesmanee, W. (2009). Effect of heat treatment on antioxidant properties of tom-kha paste and herbs/spices used in tom-kha paste. *Natural Science*, *43*, 305–312.

Bhardwaj, A.K., Kumar, R., Kumar, S., Goel, B., & Chauhan, R. (2021). Energy and exergy analyses of drying medicinal herb in a novel forced convection solar dryer integrated with SHSM and PCM. *Sustainable Energy Technologies and Assessments*, *45*. https://doi.org/10.1016/j.seta.2021.101119.

Choo, C.O., Chua, B.L., Figiel, A., Jałoszyński, K., Wojdyło, A., Szumny, A., Łyczko, J., & Chong, C.H. (2020). Hybrid drying of *murraya koenigii* leaves: Energy consumption, antioxidant capacity, profiling of volatile compounds and quality studies. *Processes*, *8*, 240. https://doi.org/10.3390/pr8020240.

Chou, P.T., & Khan, A.U. (1983). L-ascorbic acid quenching of singlet delta molecular oxygen in aqueous media: Generalized antioxidant property of vitamin C. *Biochemical and Biophysical Research Communications*, *115*(3), 932–937.

Chua, L.Y.W., Chua, B.L., Figiel, A., Chong, C.H., Wojdyło, A., Szumny, A., & Choong, T.S.W. (2019). Antioxidant activity, and volatile and phytosterol contents of *strobilanthes crispus* dehydrated using conventional and vacuum microwave drying methods. *Molecules*, *24*(7), 1397. https://doi.org/10.3390/molecules24071397.

Davey, M.W., van Montagu, M., Sanmartin, M., & Kanellis, A., Smirnoff, N. (2000). Plant ascorbic acid: Chemistry, function, metabolism, bioavailability and effects of processing. *Journal of the Science of Food and Agriculture*, *80*(7), 825–860.

Duncan, S., & Chang, H.H. (2012). Implications of light energy on food quality and packaging selection. *Advances in Food and Nutrition Research*, *67*, 25–73. https://doi.org/10.1016/B978-0-12-394598-3.00002-2.

Ekka, J.P., Bala, K., Muthukumar, P., & Kanaujiya, D.K. (2020). Performance analysis of a forced convection mixed mode horizontal solar cabinet dryer for drying of black ginger (*Kaempferia parviflora*) using two successive air mass flow rates. *Renewable Energy*, *152*, 55–66. https://doi.org/10.1016/j.renene.2020.01.035.

Fatouh, M., Metwally, M.N., Helali, A.B., & Shedid, M.H. (2006). Herbs drying using a heat pump dryer. *Energy Conversion and Management*, *47*(15–16), 2629–2643.

Gorinstein, S., Leontowicz, H., Leontowicz, M., Namiesnik, J., Najman, K., Drzewiecki, J., Cvikrová, M., Martincová, O., Katrich, E., & Trakhtenberg S. (2008). Comparison of the main bioactive compounds and antioxidant activities in garlic and white and red onions after treatment protocols. *Journal of Agricultural and Food Chemistry*, *56*(12), 4418–4426.

Harun, N.H., Abdul Aziz, A., Wan Zamri, W.M., Rahman, R.A., & Aziz, R. (2015). Optimization of process parameters for spray drying of Tongkat Ali extract. *Journal of Engineering Science and Technology*, *6*(1), 31–41.

Hii, C.L., Jangam, S.V., Ong, S.P., Show, P.L., & Mujumdar, A.S. (2015). *Processing of Foods, Vegetables, and Fruits: Recent Advances*. Retrieved from https://arunmujumdar.com/ebooks.

Hii, C.L., Ong, S.P., Yap, J.Y. Putranto, A., & Mangindaan, D. (2021). Hybrid drying of food and bioproducts: A review. *Drying Technology*, *39*(11), 1554–1576, https://doi.org/10.1080/07373937.2021.1914078

Jiménez-Monreal, A., García-Diz, L., Martínez-Tomé, M., Mariscal, M., & Murcia, M. (2009). Influence of cooking methods on antioxidant activity of vegetables. *Journal of Food Science*, *74*(3), H97–H103.

Jung, M., Kim, S., & Kim, S. (1995). Riboflavin-sensitized photooxidation of ascorbic acid: Kinetics and amino acid effects. *Food Chemistry*, *53*(4), 397–403.

Kantrong, H., Tansakul, A., & Mittal, G.S. (2014). Drying characteristics and quality of shiitake mushroom undergoing microwave-vacuum drying and microwave-vacuum combined with infrared drying. *Journal of Food Science and Technology*, *51*, 3594–3608. https://doi.org/10.1007/s13197-012-0888-4

Khanum, H., Sulochanamma, G., & Borse, B. B. (2013). Impact of drying coriander herb on antioxidant activity and mineral content. *Journal of Biological & Scientific Opinion*, *1*(2), 50–55. https://doi.org/10.7897/2321-6328.01203

Lakshmi, D.V.N., Muthukumar, P., & Nayak, P.K. (2021). Experimental investigations on active solar dryers integrated with thermal storage for drying of black pepper. *Renewable Energy*, *167*, 728–739. https://doi.org/10.1016/j.renene.2020.11.144.

Lee, D.J., Jangam, S., & Mujumdar, A.S. (2013). Some recent advances in drying technologies to produce particulate solids. *KONA Powder and Particle Journal*, *30*, 69–83. https://doi.org/10.14356/kona.2013010.

Lemus-Mondaca, R.A., Vega-Galver, A., Rojas, P., & Ah-Hen, K. (2016). Assessment of quality attributes and steviosides of *Stevia rebaudiana* leaves subjected to different drying methods. *Journal of Food and Nutrition Research*, *4*(11), 720–728.

Li, K., Zhang, M., Mujumdar, A.S., & Chitrakar, B. (2019). Recent developments in physical field-based drying techniques for fruits and vegetables. *Drying Technology*, *37*, 1954–1973. https://doi.org/10.1080/07373937.2018.1546733.

Maftoonazad, N., Dehghani, M.R., & Ramaswamy, H.S. (2020). Hybrid microwave-hot air tunnel drying of onion slices: Drying kinetics, energy efficiency, product rehydration, color, and flavor characteristics. *Drying Technology*, *40*(5), 966–986. https://doi.org/10.1080/07373937.2020.1841790.

Martín-Cabrejas, M.A., Aguilera, Y., Pedrosa, M.M., Cuadrado, C., Hernández, T., Díaz, S., & Esteban, R.M. (2009). The impact of dehydration process on antinutrients and protein digestibility of some legume flours. *Food Chemistry*, *114*(3), 1063–1068.

Maurya, V.K., Gothandam, K.M., Ranjan, V., Shakya, A., & Pareek, S. (2018). Effect of drying methods (microwave vacuum, freeze, hot air and sun drying) on physical, chemical and nutritional attributes of five pepper (*Capsicum annuum* var. annuum) cultivars. *Journal of the Science of Food and Agriculture*, *98*(9), 3492–3500. https://doi.org/10.1002/jsfa.8868.

Mejia-Meza, E.I., Yanez, J.A., Davies, N.M., Rasco, B., Younce, F., Remsberg, C.M., & Clary. C. (2008). Improving nutritional value of dried blueberries (*Vaccinium corymbosum* l.) combining microwave-vacuum, hot-air drying and freeze-drying technologies. *International Journal of Food Engineering*, *4*(5). https://doi.org/10.2202/1556-3758.1364.

Minh, N.P. (2019). Herbal tea production from peperomia pellucida leaf. *Plant Archives*, *19*(2), 449–451.

Mortezapour, H., Ghobadian, B., Minaei, S., Khoshtaghaza, M.H. (2012). Saffron drying with a heat pump–assisted hybrid photovoltaic–thermal solar dryer. *Drying Technology*, *30*, 560–566. https://doi.org/10.1080/07373937.2011.645261.

Motevali, A., Jafari, H., & Hashemi, J.S. (2018). Effect of infrared intensity and air temperature on exergy and energy at hybrid infrared-hot air dryer. *Chemical Industry and Chemical Engineering Quarterly*, *24*(1), 31–42. https://doi.org/10.2298/CICEQ170123015M.

Mujumdar, A.S. (Ed.). (1995). *Handbook of Industrial Drying* (2nd edition). New York: Marcel Dekker Inc.

Muller-Harvey, I. (2001). Analysis of hydrolysable tannins. *Animal Feed Science and Technology*, *91*(1–2), 3–20.

Ning, X., & Han, C. (2013). Drying characteristics and quality of Taegeuk ginseng (*Panax ginseng* C.A. Meyer) using far-infrared rays. *International Journal of Food Science & Technology*, *48*(3), 477–483. https://doi.org/10.1111/j.1365-2621.2012.03208.x.

Onwude, D.I., Hashim, N., Janius, R., Abdan, K., Chen, G., & Oladejo, A.O. (2017). Non-thermal hybrid drying of fruits and vegetables: A review of current technologies. *Innovative Food Science and Emerging Technologies*, *43*, 223–238. https://doi.org/10.1016/j.ifset.2017.08.010.

Perera, C.O., & Rahman, M.S. (1997). Heat pump dehumidifier drying of food. *Trends in Food Science and Technology*, *8*, 75–79. https://doi.org/10.1016/S0924-2244(97)01013-3.

Peter, K.V. (Ed.). (2012). *Introduction to Herbs and Spices: Definition, Trade and Applications: Handbook of Herbs and Spices*. Cambridge: Woodhead Publishing.

Pirbalouti, A.G., Mahdad, E., & Craker, L. (2013). Effects of drying methods on qualitative and quantitative properties of essential oil of two basil landraces. *Food Chemistry*, *141*(3), 2440. https://doi.org/10.1016/j.foodchem.2013.05.098.

Qu, W., Pan, Z., & Ma, H. (2010). Extraction modeling and activities of antioxidants from pomegranate marc. *Journal of Food Engineering*, *99*(1), 16–23.

Raksakantong, P., Siriamornpun, S., & Meeso, N. (2012). Effect of drying methods on volatile compounds, fatty acids and antioxidant property of Thai kaffir lime (*Citrus hystrix* DC). *International Journal of Food Science & Technology*, *47*(3), 603–612.

Ren, Z., Yu, X., Yagoub, A.E.A., Fakayode, O.A., Ma, H., Sun, Y., & Zhou, C. (2021). Combinative effect of cutting orientation and drying techniques (hot air, vacuum, freeze and catalytic infrared drying) on the physicochemical properties of ginger (*Zingiber officinale* Roscoe). *LWT*, *144*, 111238. https://doi.org/10.1016/j.lwt.2021.111238.

Romauli, N.D.M., Ambarita, H., Qadry, A., & Sihombing, H.V. (2021). Effect of drying whole and half chili pods using a solar dryer with $CaCl_2$ desiccant on quality of powder chili. *International Journal of Food Science*, 2021, Article ID 9731727. https://doi.org/10.1155/2021/9731727.

Rubinskiene, M., Viskelis, P., Dambrauskienė, E., Viskelis, J.V., & Rasa. K. (2015). Effect of drying methods on the chemical composition and colour of peppermint (*Mentha × piperita* L.) leaves. *Zemdirbyste*, *102*, 223–228. https://doi.org/10.13080/z-a.2015.102.029.

Sagar, V.R., & Kumar, P.S. (2010). Recent advances in drying and dehydration of fruits and vegetables: A review. *Journal of Food Science and Technology*, *47*, 15–26. https://doi.org/10.1007/s13197-010-0010.8.

Sankhé, D.D. (2019, March 2022). *Indian Spices 101: How to Work with Dry Spices*. Retrieved from https://www.seriouseats.com.

Selvi, K.Ç. (2020). Investigating the influence of infrared drying method on linden (*Tilia platyphyllos* scop.) leaves: Kinetics, color, projected area, modeling, total phenolic, and flavonoid content. *Plants, 9*(7), 916. https://doi.org/10.3390/plants9070916.

Sharma, K., Ko, E., Assefa, A., Ha, S., Nile, S., Lee, E., & Park, S. (2015). Temperature-dependent studies on the total phenolics, flavonoids, antioxidant activities, and sugar content in six onion varieties. *Journal of Food and Drug Analysis, 23*(2), 243–252.

Śledź, M., & Witrowa-Rajchert, D. (2012). Influence of microwave-convective drying on chlorophyll content and colour of herbs. *Acta Agrophysica, 19*(4), 865–876.

Suvarnakuta, P., Devahastin, S., & Mujumdar, A.S. (2005). Drying kinetics and β-carotene degradation in carrot undergoing different drying processes. *Journal of Food Science, 70*, 520–526.

Tezcan, D., Sabancı, S., Cevik, M., Cokgezme, O.F., & Icier, F. (2020). Infrared drying of dill leaves: Drying characteristics, temperature distributions, performance analyses and colour changes. *Food Science and Technology International, 27*(1), 32–45. https://doi.org/10.1177/1082013220929142.

Torki-Harchegani, M., Ghanbarian, D., Pirbalouti, A.G., & Sadeghi, M. (2016). Dehydration behaviour, mathematical modelling, energy efficiency and essential oil yield of peppermint leaves undergoing microwave and hot air treatments. *Renewable and Sustainable Energy Reviews, 58*, 407–418.

Vega-Gálvez, A., Lemus-Mondaca, R., Tello-Ireland, C., Miranda, M., & Yagnam, F. (2009). Kinetic study of convective drying of blueberry variety O'Neil (*Vaccinium corymbosum*). *Chilean Journal of Agricultural Research, 69*(2), 171–178.

Ye, L., El-Mesery, H. S., Ashfaq, M. M., Shi, Y., Zicheng, H., & Alshaer, W.G. (2021). Analysis of energy and specific energy requirements in various drying process of mint leaves. *Case Studies in Thermal Engineering, 26*, 101113.

Zhang, M., Chen, H., Mujumdar, A.S., Tang, J., Miao, S., & Wang, Y. (2017). Recent developments in high-quality drying of vegetables, fruits, and aquatic products. *Critical Reviews in Food Science and Nutrition, 57*, 1239–1255. https://doi.org/10.1080/10408398.2014.979280.

3 Drying Characteristics of Herbs, Spices, and Medicinal Plants

Chien Hwa Chong
University of Nottingham Malaysia

Nesa Dibagar
Wroclaw University of Environmental and Life Sciences

Rhonalyn Maulion
Batangas State University

Adam Figiel
Wroclaw University of Environmental and Life Sciences

CONTENTS

DOI: 10.1201/9781003269250-3

29

3.1 INTRODUCTION

The acceptability of food is significantly increased by the addition of herbs, spices, and medicinal plants. These biomaterials improve the shelf-life of food, delay its oxidation, and impart health-promoting properties (Kurup et al., 2020) as natural therapies (Sharma et al., 2009).

The physical and chemical properties of aromatic and medicinal plants are determined by their moisture content since water is the most significant component. The first step in many post-harvest operations is drying for the purpose of moisture content reduction to the equilibrium level that is defined for certain air relative humidity and temperature (Pandey and Pandey, 2017; Thamkaew et al., 2020). Therefore, adequate drying is required for aromatic and medicinal plants after harvesting to prolong the shelf-life, reduce the complexity of packaging and storage, and provide a diverse range of products for consumers, as well as meeting the requirements of energy-efficient, cost-effective, environment-friendly, and sustainability (Khallaf and El-Sebaii, 2022).

Drying can reduce the moisture content to a safe level where enzymatic activity can be stopped when the water activity is lower than 0.6 (Jin et al., 2017). At this water activity level, microorganisms can be prevented from growing and multiplying, which helps in increasing the shelf-life of food products. Drying is considered a complex thermal operation in which external and internal heat and mass transfer exhibit an unsteady state process in the various drying periods (e.g. constant rate and falling rate periods). The drying characteristics are affected by the product characteristics, drying technique, operational conditions, and heat sources. Drying kinetics are the changes in moisture content over time, and typically, Fick's second-law model or empirical models are used to describe it (Taheri-Garavand and Meda, 2018). It is important to be aware of the factors influencing the drying kinetics during lab-scale drying because the operating data can be compiled and linked to the product quality as a reference for industrial scaling up.

In the preservation of herbs, spices, and medicinal plants, various drying techniques have been reported such as microwave drying, infrared drying, solar drying, convective air drying, vacuum drying, thin-layer air infringement drying, spray drying, and fluidized bed drying. In most cases, traditional drying techniques are integrated to further enhance the drying efficiency instead of using only single technique. This chapter aims to present the drying kinetics and mathematical modeling aspects of some popular herbs, spices, and medicinal plants obtained from laboratory studies including those from industrial-scale drying techniques. A detailed discussion of the influence of various drying techniques on the content of bioactive compounds in functional food can be referred in Chong et al. (2021).

3.2 HERBS, SPICES, AND MEDICINAL PLANTS

Herbs, spices, and medicinal plants have been utilized for thousands of years, and the scientific information on botany, horticulture, and pharmacology associated with the growth and utilization of these plants is ever increasing. These plants are highly perishable due to their high moisture content (Kurup et al., 2020). Therefore, a suitable drying technique is necessary to prevent biomaterials wastage and increase their shelf-life. The drying techniques, operating conditions, and major findings in terms of drying kinetics, energy consumption, and mathematical modeling are presented in Table 3.1 for selected herbs, spices, and medicinal plants.

Solar dryers (SD) can be classified as direct solar dryers, indirect solar dryers, mixed mode, and hybrid solar dryers, which can be further categorized into natural and forced convection types (El-Sebaii and Shalaby, 2013). SD usually results in a faster rate of dehydration than sun drying due to the higher temperature inside the solar collector supplied to the drying chamber. Additional advantages of SD are such as it can continue drying even under rainy conditions, and biogas fuel could be used as an auxiliary energy source during cloudy/night hours (Sharma et al., 2021; Nukulwar and Tungikar, 2022) or integrated phase change materials (PCM) can be installed to collect more energy. For instance, some have attempted to combine it with convective air drying (CD) to reduce the drying duration and improve product quality. El-Sebaii and Shalaby (2013) used an indirect-mode forced convection solar dryer for drying the thymus and mint. The drying characteristics of the mint exhibited an initial transient period, constant rate period, and falling rate period. Fourteen thin-layer drying models were evaluated, and the Midilli and Kucuk model was best selected to describe the thin-layer solar drying of the mint ($R^2 = 0.9983$). However, the Page and modified Page models were found to be the best for drying the thymus ($R^2 = 0.9931$).

In an attempt to further improve the drying efficiency and product quality, Rabha et al. (2017) dried ghost chili peppers using a forced convection solar tunnel dryer. Drying time was recorded shorter at 123 hours (from 5.9% to 0.12% (d.b.)) as compared to sun drying at 193 hours. The Midilli and Kucuk model was the best-fitted model to describe the drying kinetics ($R^2 = 0.9972$). Borah et al. (2015a) also conducted drying of whole and sliced turmeric rhizomes (*Curcuma longa* L.) using a solar conduction dryer. The drying curve of sliced samples showed more uniform reduction in comparison to that of whole samples. The average effective moisture diffusivity was found to be 1.852×10^{-10} m^2/s for slab samples and 1.456×10^{-10} m^2/s for solid samples. When the moisture content was in the range of 60% (w.b.), similar drying rates were recorded for both sliced and whole turmeric rhizomes. In this study, the Page model was the best-fitted model for describing the drying kinetics of sliced ($R^2 = 0.9946$) and solid ($R^2 = 0.9911$) turmeric rhizomes. In other studies, the average diffusivity value of sliced turmeric samples was obtained at 7.45×10^{-8} m^2/s in direct solar drying (Sharma et al., 2021) and 1.667×10^{-9} m^2/s in natural-convection solar-biomass-integrated drying (Borah et al., 2015b).

Hybrid drying is a technique, which combines two or more drying mechanisms into one drying system. It has two configurations, namely, single-stage and

TABLE 3.1

Drying Characteristics of Herbs, Spices, and Medicinal Plants

Products	Drying Techniques	Process Parameters	Major Findings	References
			Herbs	
Basil leaf	MWD	PI: 0.3 W/g DT: 4 minutes SM: 10 g	MC reduced from 91.20% to 20% (w.b.).	Danso-Boateng (2013)
Ginseng root slices	CD IRD IR+CD: IR radiation (25 minutes) followed by CD	T: 50°C–70°C V: 1 m/s	The IRD drying time was less than in CD and IR+CD at the same drying temperature. Compared to CD, IR+CD shortened drying time and saved energy consumption by 35% and 18%, respectively. The optimum ginseng root drying was IR+CD at 50°C.	Pei et al. (2020)
Lemon grass extract	SPD	T: 110°C–150°C at 10°C interval FR: 280 mL/h V: 1.4 m/s Carriers: Arabic gum maltodextrin, and their mixture at 7:3	The MC of samples processed from 110°C to 150°C ranged from 13% to 9% (w.b.). A lower MC of the powder was obtained at a high temperature. The proposed optimum temperature was 130°C.	Tran and Nguyen (2018)
Lemon myrtle	VD MWD	T: 50°C, 70°C and 90°C P: 720, 960 and 1,200 W DT: 48 hours	In VD, drying time was reduced by 63% when the temperature increased to 90°C. The MC reduction in VD and MWD was 5.1%–4.8% and 0.4%–0.8%, respectively.	Saifullah et al. (2019)
Thymus and mint	FCD+SD	T: 39°C–54°C DT (SD): 34 hours DT (FCD): 5 hours	Drying showed constant and falling rate periods. DT of thymus and mint depended on the product mass and drying air. The best-fitted models were the Page model for thymus and Midilli and Kucuk for mint.	El-Sebaii and Shalaby (2013)

(Continued)

TABLE 3.1 (Continued)

Drying Characteristics of Herbs, Spices, and Medicinal Plants

Products	Drying Techniques	Process Parameters	Major Findings	References
Spearmint	CD IR+CD	T: 30°C (210 minutes) 40°C (120 minutes) 50°C (45 minutes)	The D_{eff} values ranged from 5.58×10^{-10} to 2.36×10^{-9} m²/s. IR+CD reduced EC by 45%–60% compared to CD.	Nalawade et al. (2019)
Spices				
Chili	CD IRD IR+CD	T: 50°C–70°C at 5°C interval SM: 200 g	The highest EC was at 70°C (0.95 kWh) and the lowest at 50°C (0.73 kWh). The optimal drying performance was obtained at 70°C, 65°C, and 50°C temperatures in CD, IR+CD, and IRD, respectively.	Mihindukulasuriya and Jayasuriya (2015)
Ghost chili pepper	OPD SD	T: 44°C–66°C V: 1.7 m/s SM: 200 g	Drying in the SD (123 hours) was faster than in OPD (193 hours). Only, the falling rate period was observed in SD and OPD. The best-fitted models were the Midilli and Kucuk for SD and Page and the modified Page for OPD.	Rabha et al. (2017)
Green pepper	CD CD+MW CD+MW+IR$_{per}$ CD+MW$_{per}$+IR$_{per}$ CD+IR$_{per}$	T: 65°C V = 1.8 m/s P(MW): 62 W P(IR): 240 W	CV+MW+IR$_{per}$ or CV+MW$_{per}$+IR$_{per}$ significantly shortened the DT even up to 76% as compared to CD. The CV+MW drying resulted also in the lowest EC of 3.3 kWh.	Lechtańska et al. (2015)
Turmeric rhizomes (*Curcuma longa* L.)	SD	T: 39°C–51°C DT: 12 hours	MC was reduced from 78.65% to 6.36% and 5.50% (w.b.) for solid and sliced samples. D_{eff} was obtained 1.85×10^{-10} and 1.46×10^{-10} m²/s for slab and solid samples, respectively. The best-fit model was Page.	Borah et al. (2015a)

(Continued)

TABLE 3.1 (Continued)
Drying Characteristics of Herbs, Spices, and Medicinal Plants

Products	Drying Techniques	Process Parameters	Major Findings	References
			Medicinal Plants	
Achillea collina, Solidago gigantean, Wormwood, Wallnut leaf, Daucus carota	FBD	V: 0–1 m/s T: 27.4°C–31.4°C	The average DR of Achillea collina and Daucus carota were the highest, followed by Solidago gigantea, walnut leaf, and wormwood.	Poós and Varju (2017)
Hyssop (*Hyssopus officinalis* L.)	OPD MWD IRD	T (OPD): 25°C–35°C MWP: 300, 450 and 600 W T (IRD): 40°C, 50°C and 60°C	IRD for hyssop drying reduced the relative DT and provided a uniform drying of the herbal substance.	Saeidi et al. (2020)
Murta (*Ugni molinae Turcz*) berries	CD IR+CD	T: 40°C, 50°C and 60°C P: 400 and 800 W	D_{eff} was obtained between 7.59×10^{-10} and 44.18×10^{-10} m²/s for CD and 11.34×10^{-10} and 85.41×10^{-10} m²/s for IR+CD. Minimum EC was achieved at 60°C and 400 W.	Puente-Díaz et al. (2013)
Murraya koenigii leaves	CD CD+MWD	P: 240 W (44 minutes), 360 W (27 minutes) and 480 W (18 minutes)	DR was increased by 91% in CD+MWD.	Choo and Chua (2019)
Peppermint leaves	CD IRD	T: 30°C, 40°C and 50°C V: 0.5, 1 and 1.5 m/s PI: 500, 3,000, and 4,000 W/m² Emitter distance: 0, 15, and 20 cm	Activation energies for IRD and CD were ranged from 0.206 to 0.439 and from 21.476 to 27.784 kJ/mol, respectively. D_{eff} ranged from 1.096×10^{-11} to 2.486×10^{-11} m²/s and 3.312×10^{-11} to 5.928×10^{-11} m²/s for CD and IRD, respectively.	Miraei Ashtiani et al. (2017)

(Continued)

TABLE 3.1 (Continued)
Drying Characteristics of Herbs, Spices, and Medicinal Plants

Products	Drying Techniques	Process Parameters	Major Findings	References
Rose petals	IRD	T: 50°C (1,680 seconds), 60°C (1,080 seconds), and 70°C (600 seconds)	MC was reduced from 86% to 12% (w.b.)	Selvi et al. (2020)
Rosella (*Hibiscus sabdariffa*) flower	SDDGC	RH: 60%–80% T: 48°C	MC was reduced from 90.84% to 76.67% in 1 day and 7.67% in 2 days. SMER reduced from 0.222 kg/kWh in 1 day and 0.0256 kg/kWh in 2 days.	Marnoto (2014)
Lavender flowers	VMD CD+VD	P: 240 W (44 minutes), 360 W (36 minutes), and 480 W (61 minutes)	CD was very effective at the beginning of the drying process and VMD at the final stage of drying. The best-fitted model was Page.	Łyczko et al. (2019)
Cassia alata	CD VMD CPD+VMFD	PI: 6, 9, and 12 W/g VP: 4–6kPa T: 50°C (90 minutes) followed by VMD (9 W/g)	In CPD+VMFD, CPD was effective in removing moisture during the initial drying process whilst VMFD assisted in the removal of residual moisture. CPD+VMFD reduced the final EC by 32.82% (kJ/$g_{f.w.}$) and 26.12% (kJ/g_w) compared with CD at 50°C. The best-fitted model was the modified Page.	Chua et al. (2019)
Strobilanthes crispus	CD VMD CPD+VMFD	PI: 6, 9 and 12 W/g VP: 4–6kPa T: 50°C (90 minutes) followed by VMD (9 W/g)	VMD at 6, 9, and 12 W/g required shorter drying durations of 28, 21, and 14 minutes, respectively. The lengthy drying process of CD was shortened to 105 minutes by introducing VMD. The best-fitted model was the modified Page.	Chua et al. (2019)

(Continued)

TABLE 3.1 (Continued)
Drying Characteristics of Herbs, Spices, and Medicinal Plants

Products	Drying Techniques	Process Parameters	Major Findings	References
Phyla nodiflora leaves	CD VMD CPD+VMFD	T: 40°C, 50°C, and 60°C Power: 6, 9 and 12 W/g VP: 4–6 kPa T: 50°C (90 minutes) followed by VMD (9 W/g)	VMD achieved the shortest drying time, whereas CD was the most time-consuming. CPD+VMFD reduced the SEC by 80.2% and 51.1% (kJ/g_w) compared with CD and VMD, respectively, at 50°C. CPD was effective in reducing water content by approximately 92.4%. The best-fitted model was the modified Page.	Chua et al. (2019)
Rosmarinus officinalis L.	CD VMD CPD+VMFD	T: 50°C, 60°C and 70°C P: 240, 360 and 480 W VP: 4–6 kPa T: 50°C (30, 60 and 120 minutes) followed by VMD finish drying (360 W)	Complete moisture removal was achieved the fastest by VMD at the highest wattage. Increasing the microwave power from 240 to 480 W resulted in a substantial increase in the DR by 75%. VMD at 480 W was the most efficient technique with the least EC. The best-fitted model was the modified Page.	Ali et al. (2020)
Murraya koenigii leaves	CD VMD CPD+VMFD	T: 40°C, 50°C and 60°C PI: 6, 9 and 12 W/g VP: 4–6 kPa	MVD as the finishing-drying technique reduced the SEC by 67.3% (kJ/$g_{r.w.}$) and 48.9% (kJ/g_w) in comparison to CD and VMD at 50°C. With VMD at 9 W/g combined with CD at 50°C, the EC was successfully reduced from 274.86 and 372.92 kJ/g_w to 84.11 kJ/$g_{f.w.}$ and 161.60 kJ/g_w respectively. The best-fitted model was the modified Page.	Choo et al. (2020)
Citrus hystrix leaves	CD VMD CPD+VMFD FD	T: 50°C (15 minutes interval until falling rate period) followed by VMD finish drying (3 minutes interval) PI: 6, 9 and 12 W/g VP: 4–6 kPa P: 65 Pa, T: −60°C for FD	CPD+VMFD lowered the specific energy consumption by 51.6%; meanwhile, the CMV set at 12 W/g retained the highest amount of antioxidant, TPC, and volatile compounds. The best-fitted model was the modified Page.	Choo et al. (2022)

two-stage hybrid drying systems. In the single-stage system, two drying techniques are combined, while in the two-stage system, pre-drying and finishing-drying stages are included. For example, infrared-assisted convective drying (CV+IR$_{per}$) was used by Łechta-ska et al. (2015) to dry green pepper. Hot air in combination with periodic infrared radiation shortened the drying time and reduced the total energy consumption by 38% and 6.9%, respectively, as compared to convective drying alone (CD). CV+IR$_{per}$ rapidly increased the drying rate at the beginning of the process and used only convective drying in the subsequent stages. It was also noted that the efficiency of infrared drying (IRD) reduced as drying progressed as it focused more on surface moisture removal and could not effectively penetrate the product. In addition, Mihindukulasuriya and Jayasuriya (2015) attempted to dry chili using a combined infrared and hot air rotary dryer. No constant rate period was observed due to the rapid heat and mass transfer across the product. Additionally, the time needed to reduce the moisture ratio of the product during the combined infrared and hot air mode (2 hours) was lower than that of hot air drying (5 hours) alone.

IRD has been reported to be used for drying of medicinal plants. Energy from IR radiation directly heats the sample without a transfer medium and delivers heat to the material being dried. The delivered energy then leads to internal heating, which heats the whole material and the internal moisture (Delfiya et al., 2022). Selvi et al. (2020) investigated the effect of IR drying on Rose petals, and the drying time reduced significantly with the increase in IR temperature. The drying time reduced from 1,680 to 600 seconds when the IR temperature increased from 50°C to 70°C. The effect of different drying techniques including IRD on the drying characteristics of hyssop (*Hyssopus officinalis* L.) has also been investigated by Saeidi et al. (2020). Hyssop samples were dried in 367 minutes using IRD, which was longer than microwave drying (20 minutes) but shorter than oven drying (1,997 minutes). Miraei Ashtiani et al. (2017) analyzed the drying characteristics of peppermint leaves under IRD treatments and found that the moisture ratio of the leaves decreased much more rapidly when the IR radiation intensity increased. This resulted in an increase in the material temperature and vapor pressure and eventually its moisture removal rate (Delfiya et al., 2022).

Łyczko et al. (2019) conducted vacuum-microwave drying (VMD) on lavender (*Lavandula angustifolia* Mill.) flowers, and it was found that the Page model could describe well the drying kinetics ($R^2 > 0.99$). The drying time was reduced from 44 to 20 minutes when the microwave power increased from 240 to 480 W. The drying time for VMD was shorter as compared to CD and convective pre-drying with vacuum-microwave finish drying (CPD+VMFD), reaching a total drying time of 44, 27, and 18 minutes for 240, 360, and 480 W, respectively. This significant reduction in the total drying time of VMD is mainly due to moisture diffusion that is supported by a pressure diffusion mechanism of the Darcy type (Delfiya et al., 2022). The drying rate for VMD was also the highest among all the drying techniques and showed a higher drying rate at the initial stage and a lower drying rate when toward the falling rate period. From these studies, VMD was found to have a shorter drying time as compared to CD with an increase in microwave power output.

3.3 SUCCESSFUL EXAMPLES ON INDUSTRIAL-SCALE DRYING

Many industrial-scale drying technologies have been developed for various herbs, spices, and medicinal plants. However, most of the technologies still adopt conventional drying techniques and only very few use hybrid drying techniques. This section will discuss the design of some industrial-scale dryers and the associated drying characteristics.

3.3.1 INDUSTRIAL-SCALE DRYING OF SPICES

3.3.1.1 Chili

The storage of solar energy as heat is possible using PCM, which is characterized by high volumetric storage densities and small temperature intervals for energy extraction and regeneration. PCMs are classified into various categories with organic and inorganic compounds and eutectic mixtures of these compounds at various phase-transition temperatures (Huang et al., 2019). Pankaew et al. (2020) investigated the performance of a greenhouse solar dryer integrated with PCM as latent heat storage for chili drying. About 500 kg chili with an initial moisture content of 74.7% (w.b.) was dried to 10% (w.b.) in 2.5, 3.5, and 11 days, respectively, using the solar dryer with PCM thermal storage, solar dryer without PCM thermal storage, and open sun drying. The exergy efficiencies of the solar dryer with and without the PCM thermal storage were determined at 13.1% and 11.4%, respectively.

Kumar et al. (2020) tested a natural convection solar greenhouse dryer for drying of 250 kg red chili. The dryer reduced the moisture content of red chili from 79% (w.b.) to about 10% (w.b.) in 55 hours as compared to 124 hours by open sun drying (about 56% saving in drying time). The thermal efficiency of the dryer was determined at 16.25% with specific energy consumption of 6.06 kWh/kg. In a different study, Kaewkiew et al. (2012) investigated the performance of a greenhouse-type solar dryer to dry between 300 and 1,000 kg of chili. The dryer had a parabolic shape covered with polycarbonate sheets, and the base of the dryer was a concrete floor measuring $8 \times 20 \, m^2$ area. It was constructed with nine ventilation fans and three 50 W solar cell modules. On an Average, it was found that 500 kg of chili with an initial moisture content of 74% (w.b.) were dried to a final value of 9% (w.b.) within 3 days, while sun drying required 5 days.

Janjai et al. (2011) reported drying of chili and coffee beans using a solar greenhouse dryer with a maximum product capacity of 1,000 kg. About 300 kg of chili with an initial moisture content of 75% (w.b.) was dried to 15% (w.b.) within 3 days while sun drying required 5 days. Using the same dryer, 200 kg of coffee with an initial moisture content of 52% (w.b.) was dried to 13% (w.b.) within 2 days as compared to sun drying that required 4 days. Process economics showed that the payback period of the dryer was estimated at 2.5 years.

ELkhadraoui et al. (2015) investigated the performance of a mixed-mode solar greenhouse dryer with forced convection to dry 80 kg red pepper. The dryer consisted of a flat-plate solar collector and a chapel-shaped greenhouse structure. Red peppers with initial moisture content of 12.15% (d.b.) were dried to 0.17% (d.b.) in the solar greenhouse solar dryer within 17 hours as compared to sun drying that required about 24 hours to dry to 0.19% (d.b.). This is mainly attributed to the much higher drying rate in the solar greenhouse dryer as compared to a typical solar and sun dryer. The payback period was determined at 1.6 years which was much less than the estimated life span (20 years).

Hossain and Bala (2007) investigated a mixed-mode-type forced convection solar tunnel dryer to dry red and green chilies under the tropical weather conditions of Bangladesh. The dryer consisted of a transparent plastic-covered flat-plate collector and a drying tunnel connected in series to supply hot air directly into the tunnel using two blowers operated by a photovoltaic module. The dryer had a loading capacity of 80 kg of fresh chilies. The moisture content of the red chili reduced from 285% to 50% (d.b.) in 20 hours in a solar tunnel dryer, but it took 32 hours to reduce the moisture content to 90% (d.b.) in the improved sun drying and 40% (d.b.) in the conventional sun drying, respectively. In the drying green chili, solar tunnel drying required 22 hours to reduce the moisture content from 760% to 60% (d.b.), but it took 35 hours to reach 10% and 70% (d.b.) in the improved and conventional sun drying, respectively. A combination of blanching and solar tunnel drying led to a considerable reduction in drying time as compared to conventional sun drying. The average air temperature increment in the dryer was about 22°C above ambient temperature.

An indirect two-stage solar tunnel dryer was developed by Getahun et al. (2021) to study the drying characteristics of two chili pepper varieties (*Bako Local* (BL) and *Mareko Fana* (MF)) at 45, 55 and 65 kg. It was observed that the drying times required to reach the safe storage moisture content of 11%–12% (w.b.) were 50, 70, and 75 hours for chili layer densities of 5.88, 7.19, and 8.50 kg/m², respectively, for the MF variety, whereas it took about 60, 65, and 80 hours for the BL variety. The moisture diffusivity of both varieties varied from 2.94×10^{-9} to 4.47×10^{-9} m²/s, and the efficiencies of the collectors ranged from 66.44% to 76.53%. The maximum drying rate of MF and BL varieties was 0.325 and 0.24 kg/kg·h in a thin-layer chili density. The overall system efficiency was between 24% and 31% with specific energy consumption between 2.1 and 2.44 kWh/kg, and the CO_2 emission was determined at 53.8 kg/kWh. The carbon footprint mitigation of the solar dryer showed that the renewable energy-based drying system offered better benefits with the maximum and minimum earned carbon credit of $162.880 and $136.250 per year, respectively.

3.3.1.2 Cayenne Pepper

Hempattarasuwan et al. (2019) tested a parabolic greenhouse-type solar dryer for drying of cayenne pepper (*Capsicum annuum*) with the loading capacity of 100–200 kg. It had a parabolic roof structure covered with polycarbonate sheets placed on a concrete floor and used three DC fans powered by a 50-W solar cell module. The moisture content of the whole pods reduced within 19 hours from 72% to 36% (w.b.) and 8% (w.b.) in the solar dryer and tray dryer, respectively. On the other hand, the cut pods dried within 7 hours from an initial moisture content of 70% to 22% (w.b.) and 4% (w.b.) in the solar dryer and tray dryer, respectively.

3.3.1.3 Ginger and Turmeric

Nimnuan and Nabnean (2020) assessed the performance of an industrial-scale greenhouse-type solar dryer for drying of cassumunar ginger (*Plai*). The dryer consisted of a parabolic roof structure with polycarbonates covering, and the base of the dryer was a concrete floor with area measuring 9 m ×12.4 m. Nine DC fans powered by three 50 W PV modules were used to ventilate the dryer. About 300 kg of cassumunar ginger was dried at air temperatures varied from 30°C to 55°C during drying. The drying time in the solar greenhouse dryer was 1 day to dry the cassumunar ginger from 90% to 10% (w.b.) as compared to natural sun drying which dried the products only to 40% (w.b.) within the same duration. The solar greenhouse dryer enabled a 67% reduction in cassumunar ginger drying time as compared to sun drying with an average drying efficiency of 38.9%.

Borah et al. (2015b) utilized a solar-biomass integrated batch drying (IDS) system with a capacity of 100 kg/batch to dry ginger and turmeric rhizomes (600 kg). A parabolic solar collector coupled with bio-waste-fired combustion and heating was constructed and attached to a drying chamber consisting of six trays and a wind turbine. Fluidized bed dryer (FBD), electrical oven dryer (EO), and open sun dryer (OPD) were compared. Effective moisture diffusivity in turmeric drying was nearly 21% higher than that in ginger drying. Minimum specific energy consumption (SEC) was observed in IDS, and it was 14 and 30 times lower than FBD and OSD, respectively. The IDS also showed 36% of overall energy utilization efficiency.

3.3.1.4 Garlic

Yadav (2017) investigated the dying efficiency and performance of solar tunnel drying of garlic. The dryer was a tunnel-like structure in semi-cylindrical shape and covered with a UV-stabilized polyethylene sheet. The capacity tested was 400 kg of garlic under controlled environmental conditions. The temperature inside the dryer was always higher at 8°C–30°C above the ambient temperature. The initial moisture content of the garlic was reduced from 66% to about 9% (w.b.) in 9 days. The average thermal efficiency of the dryer was determined at 14%.

Aritesty and Wulandani (2014) used a greenhouse-effect solar rack-type dryer to dry 21 and 60 kg of sliced ginger. The average drying rate of sliced wild ginger was 17.2 kg water/kg d.m. h to dry 21 and 60 kg of wild ginger from 80% (w.b.) to 8%–11% (w.b.) and required total drying time of 27.5 and 30 hours, respectively. The best drying performance was drying of 60 kg sliced wild ginger at 47.2°C for 30 hours which was represented by a drying efficiency of 8% and total energy consumption of 29 MJ/kg.

3.3.1.5 Onion

Jain (2007) developed a transient analytical model to study a solar crop dryer having reversed absorber plate-type collector and thermal storage with natural airflow (Figure 3.1). The performance of the crop dryer (1 m² area) with a packed bed and airflow channel was evaluated for drying of 95 kg onions. A 30°-inclined absorber plate with thermal storage and 0.12 m width of airflow channel induced the mass flow rate in the range of 0.032–0.046 kg/s during drying. It was found that the crop moisture content reduced from 614% to 27% (d.b.) in 24 hours. However, the moisture content remains relatively constant after 16 hours of drying. The drying rate changed from 0.4 to 0.35 kg water/kg dm·h within 6 hours. The drying rate dropped sharply from about 0.35 to 0.05 kg water/kg dm·h for the next 10 hours. The thermal energy storage affected drying during the off-sunshine hours and was very pertinent in reducing the fluctuation in temperature for drying.

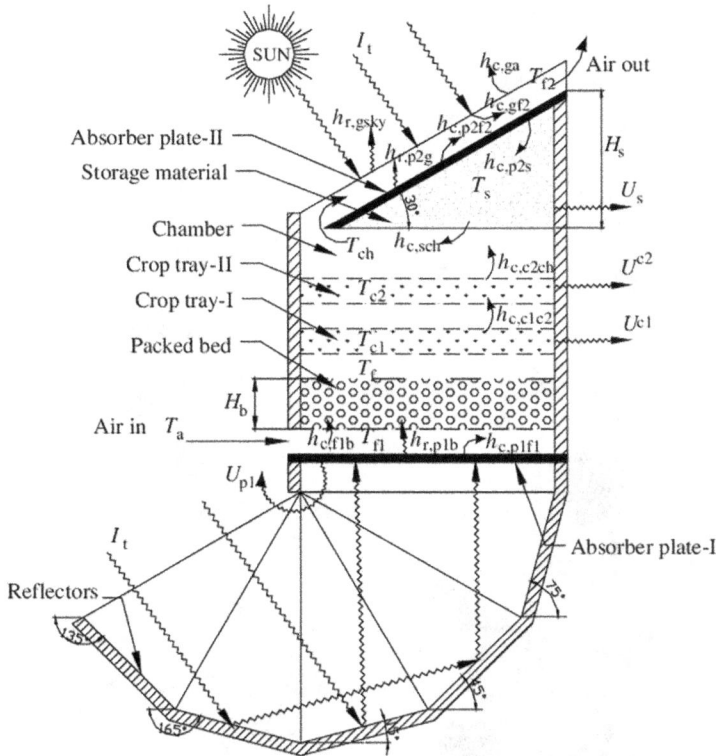

FIGURE 3.1 Schematic view of the reversed absorber with thermal storage natural convective solar crop dryer. (Adapted from Jain, 2007.)

3.3.1.6 Cardamom

An indirect active solar-biomass hybrid dryer (SBHD) (Figure 3.2) was developed by Shreelavaniya et al. (2021). The experiments were carried out using a hybrid (solar with biomass) and biomass mode to assess the performance of the dryer. Freshly harvested small cardamom (20 kg) with a moisture content of 82.4% (w.b.) was dried to 9.1% (w.b.) in hybrid mode. In biomass mode, it was able to reduce the moisture content from 83.2% to 9.5% (w.b.). The drying period of 19 hours was sufficient in both modes of drying to attain the required final moisture. The average drying efficiency of 29% was observed in the SBHD while 24% was observed from only biomass mode.

Mandal et al. (2018) developed a low-cost portable biomass-fired dryer for cardamom. The dryer had two sections that consisted of the drying trays in the upper half and the biomass burner in the lower half. The upper half and lower half weighed 22 and 16 kg, respectively, making them lightweight and portable. It was observed that large cardamom was dried from about 80% to 10% (w.b.) final moisture content in 14 hours for 20 kg loading and 19 hours for 30 kg loading. It was also found that for drying 20 and 30 kg of fresh cardamom, 14 and 19.5 kg of biomass were burnt and the

FIGURE 3.2   Solar-biomass hybrid dryer developed for small cardamom drying. (Adapted from Shreelavaniya et al., 2021.)

moisture removed was 14.07 and 20.8 kg, respectively. Thus, the average fuel requirement was 0.67 kg/kg for large cardamom. The cost of the dryer was $127.66, and the cost of drying 100 kg of cardamom was $3.81.

3.3.2   INDUSTRIAL-SCALE DRYING OF HERBS AND MEDICINAL PLANTS

3.3.2.1   Amaranth Leaves

Romuli et al. (2019) investigated the performance of an inflatable solar dryer (ISD) to dry 200 kg of amaranth leaves (*Amaranthus* spp.). In order to handle the drying of lightweight materials, the modification was made by adding an air deflector and trays inside the ISD. The air deflector integrated into the pre-heating zone helped to avoid the dried leaves from being blown away during drying. The estimated air mass flow in the inlet of the ISD was 0.75 kg/s. The total drying time was 68 hours, whereas the final moisture content of the leaves dried in the ISD and sun drying was 11.6% and 24.7% (w.b.), respectively. It was recommended to remove the stalks and chopped the leaves prior to drying to increase the ISD performance to increase the effective diffusivity of the plant samples using smaller particle sizes (Chong e al., 2019).

3.3.2.2 Chamomile

Amer and Gottschalk (2012) investigated solar drying of chamomile with energy storage using a water tank system. It consisted of a flat-plate collector, reflector, drying unit, an auxiliary electric heater, and a water tank (Figure 3.3). The solar system was tested during the summer season in Germany at a drying capacity of 32–35 kg. Drying was conducted at 18–30 hours to reduce the moisture contents of chamomile from 72%–78% to 7.9% (w.b.) and 33–48 hours to reduce it from 72%–78% to 10% (w.b.) by the solar dryer and natural sun dryer, respectively. The dryer was also able to be used with an auxiliary heat source under adverse weather conditions. By using the water tank with the solar dryer, about 15°C–20°C above ambient temperature can be stored in water during the time of sunshine. At night, the system transferred the stored heat from the water to the air inside the solar dryer to continue drying.

3.3.2.3 Red Jujube

Yuan et al. (2019) used a partially open-loop heat pump dryer with a unit room (HPDU) for an industrial-scale drying of red jujube (2,590 kg) (Figure 3.4). The unit room was designed to enable the ambient air to be mixed with the return air, thereby reducing the influence of the ambient air on the system performance, while maintaining a high system thermal efficiency. The energy benefit of the proposed HPDU was compared with a closed-loop heat pump dryer (CHPD). It was evident that a maximal specific moisture extraction rate (SMER) and minimal total energy consumption (TEC) existed when changing the bypass factor of the HPDU under certain ambient temperatures. Compared to the CHPD, the coefficient of performance (COP) of the HPDU increased by up to 40%, presenting a significant energy

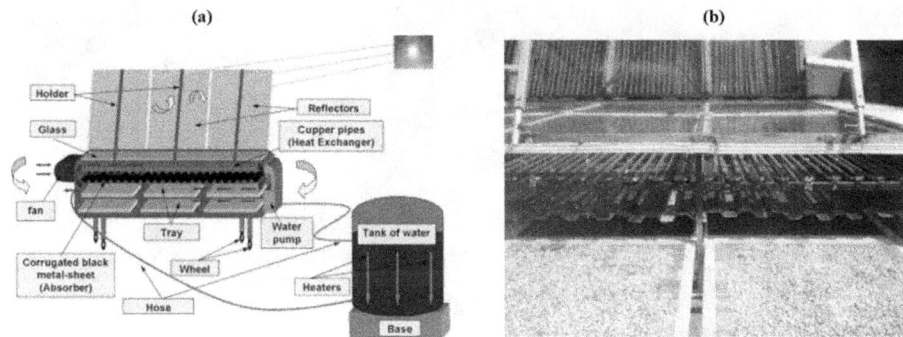

FIGURE 3.3 (a) Schematic diagram of a solar hybrid dryer with a water tank; (b) fresh chamomile put over tray in the solar dryer. (Adapted from Amer and Gottschalk, 2012.)

benefit for the application of HPDU. TEC of HPDU increased and decreased by 45% and 31%, respectively. By using the optimal bypass factor, the corresponding coefficient of performance (COP) could reach 2.79 maximum for the HPDU, compared to only 2.0 in the CHPD. Even though the ambient temperature had a considerable influence on the SMER and TEC of the system, it only slightly affected the system's COP. It had demonstrated that the HPDU outperformed the CHPD in terms of energy efficiency.

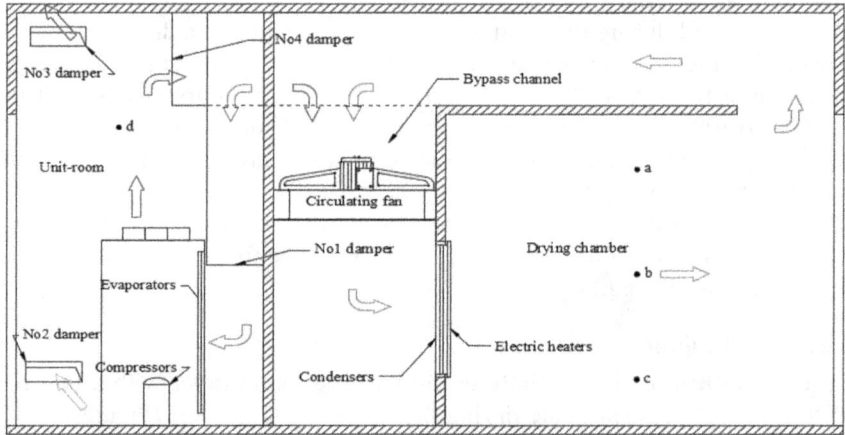

(a)

(b)

FIGURE 3.4   The novel partially open-loop heat pump dryer with the unit room: (a) schematic diagram; (b) practical prototype. (Adapted from Yuan et al., 2019.)

3.3.2.4 Mint

A hybrid photovoltaic thermal (PVT) ultraviolet (UV)-stabilized polyethylene greenhouse dryer was developed by Nayak et al. (2011) to lower the initial costs of industrial-scale drying of 100 kg of mint leaves. The drying process from an initial moisture content of 80% (w.b.) to a final moisture content of 11% (w.b.) took 21 hours, and the final weight of the dried sample was 8.4 kg. Therefore, a total of 38.7 kg of moisture was evaporated from drying 47.1 kg of mint leaves using the dryer. The results show that the efficiency of the dryer and net CO_2 mitigation over the lifetime were 34.2% and 140.97 tons, respectively. The carbon credit earned ranged from a minimum of $704.85 to a maximum of $2,819.40 per ton of carbon. Net mitigation over the lifetime was calculated as 140.97 tons.

3.3.2.5 Spearmint Leaves

Nour-Eddine et al. (2013) developed an indirect forced convection solar dryer in Algeria that consisted of an air flat-plate collector and a drying cabin to dry 20 kg spearmint leaves. It was found that the drying kinetics of the spearmint leaves showed no constant rate period. The diffusion coefficient of spearmint leaves at drying temperature within 40°C–60°C was 1.9871×10^{-11} to $6.935 \times 10^{-11} m^2/s$. In addition, the solar fraction reached its maximum values (60%–85%) at the low air mass flow rates. For a fresh product weight of 10 kg and a collector area of $2 m^2$, the solar fraction and the efficiency of the solar collector are about 60% and 50%, respectively.

3.3.2.6 Roselle

Kareem et al. (2017) investigated the performance of a forced convective multi-pass solar air heating collector (MPSAHC) system assisted with granite as a sensible energy-storing matrix. The investigation was conducted under the daily average relative humidity, solar irradiance, ambient temperature, and wind speed of 65%, 637 W/m², 32.24°C, and 0.81 m/s, respectively. About 75.2 kg of Roselle was dried from 85.6% to 9.2% (w.b.) in 14 hours. An average drying rate of 34 g/(kg.m².h) was achieved while the system optical efficiency, collector efficiency, drying efficiency, and moisture pickup efficiency of 71%, 64%, 36%, and 67% were obtained, respectively. The average drying rate of 12.43 g/(kg.m².h) was achieved for open sun drying (OSD). The drying operation in the MPSAHC system was completed within 14 hours, but it took 35 hours to complete the OSD process. The MPSAHC dryer showed comparable color retention to those from OSD under the same weather condition. The techno-economic analysis showed a payback period of 2.14 years.

Janjai et al. (2008) assessed the performance of solar drying of Roselle flower using a roof-integrated solar dryer (Figure 3.5). The dryer was used to dry 200 kg of fresh Roselle flower from a moisture content of 90% (w.b.) to a moisture content of 18% (w.b.) within 3 days. The roof-integrated SD resulted in a considerable reduction in drying time as compared to sun drying, and the solar-dried products showed a well-accepted quality. The payback period of the roof-integrated solar dryer was about 5 years.

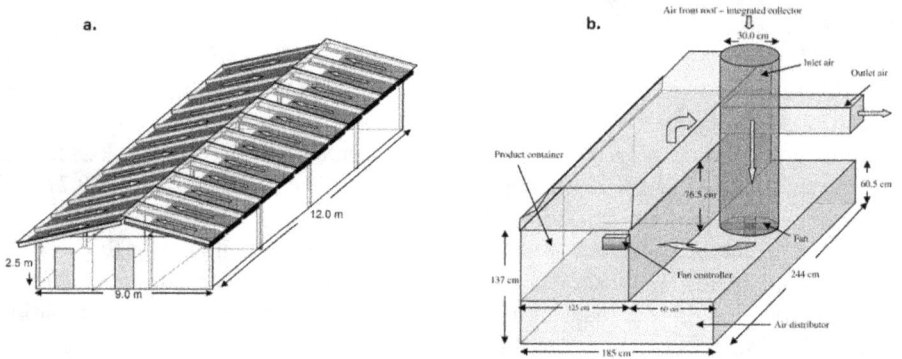

FIGURE 3.5   (a) Roof-integrated solar drying system; (b) the drying bin. (Adapted from Janjai et al., 2008.)

3.3.2.7   White Turmeric

Lakshmi et al. (2019) investigated the energy analysis and drying kinetics of *Curcuma zedoaria* (white turmeric) dried in mixed and indirect active parallel flow SD. Drying time was about 60% shorter than OSD with an initial moisture content of 318.26% (d.b.) and dried to final moisture content of 9.25% (d.b.). The overall dryer efficiency of the mixed-mode solar dryer (MFSCD) was 26.5% higher than the indirect-mode (IFSCD) solar dryer. The effective moisture diffusivity of the product dried using MFSCD (4.05×10^{-10} m²/s) was much higher as compared to IFSCD (3.14×10^{-10} m²/s).

3.3.2.8   Basil Leaves

A solar tunnel dryer was assessed by Badgujar et al. (2018) for Basil leaves drying (25 kg). The maximum temperature of 61°C was recorded at noon time (at 1,330 hours) in the solar tunnel dryer, which was 41% higher than the maximum ambient temperature. An average total drying time of 12.5 hours was required to reduce the moisture content of the leaves from 1011.11% (d.b.) to a final moisture content of 33.33% (d.b.) while OSD required an average of 15.5 drying hours to achieve the same final moisture content. The net present worth, benefit–cost ratio, and payback period for the solar tunnel dryer were Rs. 62,894.23, 1.11, and 7 months 2 days, respectively.

3.3.2.9   Jew's Mallow, Spearmint, and Parsley

Fatouh et al. (2006) developed a heat pump-assisted dryer to investigate the drying characteristics of various herbs. R134a was used as a working fluid in the heat pump circuit, and drying experiments were conducted on Jew's mallow, spearmint, and parsley. Surface loads of 3.5, 7, 14, 21, and 28 kg/m² and initial moisture contents of around 81.2%–83% (w.b.) were used, respectively. A maximum dryer productivity of about 5.4 kg/m².h was obtained using air temperature of 55°C, air velocity of 2.7 m/s, and dryer surface load of 28 kg/m². Results showed that a high surface load of 28 kg/m² at the velocity of 1.2 m/s yielded the lowest drying rate (1.30 kg/m².h), while the drying air at 55°C and velocity of 2.7 m/s achieved the first (2.00 kg/m².h) and second highest (1.66 kg/m^2.h) drying rates. There was a constant rate period of about 1.66 kg water/kg dry matter·h for the small herbs, which covered about half of the total drying time. However, the large herbs showed three constant rate periods of about 1.43, 1.02, and 0.71 kg water/kg dry matter·h before transitioned into a falling rate period. A comparison of the drying characteristics of different herbs revealed that parsley required the lowest specific energy consumption (3,684 kJ/kg water) followed by spearmint (3,982 kJ/kg water) and Jew's mallow (4,029 kJ/kg water).

3.3.2.10 Vanilla Pods

Romero et al. (2014) developed an indirect solar dryer for drying 50 kg of vanilla pods which consisted of a solar absorber made of galvanized sheet metal and a chamber. A 62% reduction was obtained in vanilla weight within 1 month of drying as compared to the traditional method that required 3 months. Comparison between CFD simulation and the thermal measurements showed good similarity between the measured and predicted temperatures at the solar collector outlet. However, a larger difference was observed in the drying cabinet. This disagreement was mainly due to the assumption of a constant convection heat transfer coefficient in the surrounding condition.

3.3.2.11 Lemon Balm and Peppermint

The drying of some leafy medicinal plants was investigated using a solar drying system developed by Amer (2012) which consisted of a flat-plate collector, reflector, drying unit, auxiliary electric heater, and water tank. The solar system was tested for drying of 16–20 kg of fresh lemon balm (*Melissa officinalis L.*) and peppermint (*Mentha pepperita*). The reflected solar radiation was found at about 57% of global solar radiation, and this radiation was added to the solar collector to increase its efficiency. Drying times for lemon balm were about 8–10 hours from moisture contents of 70%–72% to 9% (w.b.) and 33 hours from 70%–72% to 17% (w.b.) using the solar dryer and natural sun dryer, respectively. However, the moisture content of peppermint was reduced from 69%–71% to 8%–9% (w.b.) in 10–12 hours and to 10% (w.b.) in 51 hours using the solar dryer and natural sun dryer, respectively. The dryer was also fitted with an auxiliary heat source for use under adverse weather conditions. Drying could be conducted at night by using electric heaters for keeping the water temperature inside the tank at 40°C during night time and during low intensity solar radiation in day time.

3.4 CONCLUSIONS, RECOMMENDATIONS, AND FUTURE DEVELOPMENT

In general, drying is the first step in post-harvest processing due to the high amount of moisture in the fresh plant parts which could further impact their physical and chemical properties. Drying is essentially defined as the process of lowering the moisture content of plants to cease enzymatic and microbiological activity with the advantage of a longer shelf-life in the final dried products. Hence, appropriate dryers are required to swiftly lower the moisture content of herbs, spices, and medicinal plants without compromising their product quality and active ingredients.

Particularly, the following criteria must be fulfilled during drying (i) moisture content must be reduced to an equilibrium level that is defined for certain relative air humidity and temperature; (ii) maximize the retention of the active ingredients, color, flavor, and aroma; (iii) microbial count must be below the established thresholds; and (iv) dryer configuration and drying parameters (particularly temperature, mass load density, and particle size) are the key parameters affecting the drying kinetics, energy consumption, and drying efficiency.

In addition, literature shows that solar drying is an appropriate technique for herbs, spices, and medicinal plants provided that the materials are properly prepared and additional energy sources are used. Although existing hybrid SD are very advanced, there is a need for greater integration of renewable energy sources to eliminate the uncertainties from daily weather fluctuations. Thus, alternative that incorporates heat source from combusting biomass should be further envisaged. Access to cheaper photovoltaic panels will allow the use of forced airflow and heat pumps in drying facilities without access to electricity supply installation. It is also possible to use vacuum, microwave, or sonication to shorten the drying time significantly. For a quick return on investment costs, it is necessary to ensure that the hybrid system is able to work efficiently for the plant materials and determine the optimal process parameters using computer-aided systems.

Further studies on energy optimization of solar drying of herbs, spices, and medicinal plants should consider the possibility of using wind and geothermal energies as well as waste heat. The implementation of the concept of sustainable use of alternative energy sources can significantly contribute to the carbon footprint mitigation and reduction of greenhouse gas emissions.

REFERENCES

Ali, A., Oon, C. C., Chua, B. L., Figiel, A., Chong, C. H., Wojdylo, A., Turkiewicz, I. P., Szumny, A., & Łyczko, J. (2020). Volatile and polyphenol composition, anti-oxidant, anti-diabetic and anti-aging properties, and drying kinetics as affected by convective and hybrid vacuum microwave drying of Rosmarinus officinalis L. *Industrial Crops and Products*, *151*, 112463. https://doi.org/10.1016/j.indcrop.2020.112463

Amer, B. (2012). Utilization of a developed solar storage and drying system for continuous dehydration of some leafy medicinal plants. *Misr Journal of Agricultural Engineering*, *29*(1), 377–394. https://doi.org/10.21608/MJAE.2012.102613

Amer, B. M. A., & Gottschalk, K. (2012). Drying of chamomile using a hybrid solar dryer. *Post Harvest, Food and Process Engineering. International Conference of Agricultural Engineering - CIGR-AgEng: Agriculture and Engineering for a Healthier Life*, Valencia, Spain, 8–12 July.

Aritesty, E., & Wulandani, D. (2014). Performance of the rack type-greenhouse effect solar dryer for wild ginger (*Curcuma xanthorizza* Roxb.) drying. *Energy Procedia*, *47*, 94–100. https://doi.org/10.1016/J.EGYPRO.2014.01.201

Badgujar, C. M., Karpe, O. S., & Kalbande, S. R. (2018). Techno-economic evaluation of solar tunnel dryer for drying of basil (*Ocimum sanctum*). *International Journal of Current Microbiology and Applied Sciences*, *7*(07), 332–339. https://doi.org/10.20546/IJCMAS.2018.707.040

Borah, A., Hazarika, K., & Khayer, S. M. (2015a). Drying kinetics of whole and sliced turmeric rhizomes (*Curcuma longa* L.) in a solar conduction dryer. *Information Processing in Agriculture*, *2*(2), 85–92. https://doi.org/10.1016/j.inpa.2015.06.002

Borah, A., Sethi, L. N., Sarkar, S., & Hazarika, K. (2015b). Effect of drying on texture and color characteristics of ginger and turmeric in a solar biomass integrated dryer. *Journal of Food Process Engineering*, *40*(1), e12310. https://doi.org/10.1111/JFPE.12310

Chong, C. H., Figiel, A., Szummy, A., Wojdyło, A., Chua, B. L., Khek, C. H., & Yuan, M. C. (2021). Herbs drying. *Aromatic Herbs in Food: Bioactive Compounds, Processing, and Applications*, 167–200. https://doi.org/10.1016/B978-0-12-822716-9.00005-6

Chong, C. H., Law, C. L., Figiel, A., & Asni, T. (2019). Diffusivity in drying of porous media. In Xu, P., Sasmito, A.P., Mujumdar, A.S. (eds) *Heat and Mass Transfer in Drying of Porous Media*, 37–54, CRC Press. https://doi.org/10.1201/9781351019224-2/

Choo, C. O., & Chua, B. L. (2019). Potential study to reduce total drying time via hybrid drying process (Vol. 2137, No. 1, p. 020001). *AIP Conference Proceedings*, *2137*(August). https://doi.org/10.1063/1.5120977

Choo, C. O., Chua, B. L., Figiel, A., Jałoszyński, K., Wojdyło, A., Szumny, A., Łyczko, J., & Chong, C. H. (2020). Hybrid drying of Murraya koenigii leaves: Energy consumption, antioxidant capacity, profiling of volatile compounds and quality studies. *Processes*, *8*(2). https://doi.org/10.3390/pr8020240

Choo, C. O., Chua, B. L., Figiel, A., Jałoszyński, K., Wojdyło, A., Szumny, A., Łyczko, J., & Chong, C. H. (2022). Specific energy consumption and quality of *Citrus hystrix* leaves treated using convective and microwave vacuum methods. *Journal of Food Processing and Preservation*, 46, e1687 3. https://doi.org/10.1111/jfpp.16873

Chua, L. Y. W., Chua, B. L., Figiel, A., Chong, C. H., Wojdyło, A., Szumny, A., & Łyczko, J. (2019). Drying of *Phyla nodiflora* leaves: Antioxidant activity, volatile and phytosterol content, energy consumption, and quality studies. *Processes*, *7*(4). https://doi.org/10.3390/pr7040210

Danso-Boateng, E. (2013). Effect of drying methods on nutrient quality of basil (*Ocimum viride*) leaves cultivated in Ghana. *International Food Research Journal*, *20*(4), 1569–1573.

Delfiya, D. S. A., Prashob, K., Murali, S., Alfiya, P. V., Samuel, M. P., & Pandiselvam, R. (2022). Drying kinetics of food materials in infrared radiation drying: A review. *Journal of Food Process Engineering*, *45*(6), e13810. https://doi.org/10.1111/JFPE.13810

El-Sebaii, A. A., & Shalaby, S. M. (2013). Experimental investigation of an indirect-mode forced convection solar dryer for drying thymus and mint. *Energy Conversion and Management*, *74*, 109–116. https://doi.org/10.1016/j.enconman.2013.05.006

ELkhadraoui, A., Kooli, S., Hamdi, I., & Farhat, A. (2015). Experimental investigation and economic evaluation of a new mixed-mode solar greenhouse dryer for drying of red pepper and grape. *Renewable Energy*, *77*, 1–8. https://doi.org/10.1016/J.RENENE.2014.11.090

Fatouh, M., Metwally, M. N., Helali, A. B., & Shedid, M. H. (2006). Herbs drying using a heat pump dryer. *Energy Conversion and Management*, *47*(15–16), 2629–2643. https://doi.org/10.1016/J.ENCONMAN.2005.10.022

Getahun, E., Gabbiye, N., Delele, M. A., Fanta, S. W., & Vanierschot, M. (2021). Two-stage solar tunnel chili drying: Drying characteristics, performance, product quality, and carbon footprint analysis. *Solar Energy*, *230*, 73–90. https://doi.org/10.1016/J.SOLENER.2021.10.016

Hempattarasuwan, P., Somsong, P., Duangmal, K., Jaskulski, M., Adamiec, J., & Srzednicki, G. (2019). Performance evaluation of parabolic greenhouse-type solar dryer used for drying of cayenne pepper. *Drying Technology*, *38*(1–2), 48–54. https://doi.org/10.1080/07373937.2019.1609495

Hossain, M. A., & Bala, B. K. (2007). Drying of hot chilli using solar tunnel drier. *Solar Energy*, *81*(1), 85–92. https://doi.org/10.1016/J.SOLENER.2006.06.008

Huang, C. C., Wu, J. S. B., Wu, J. S., & Ting, Y. (2019). Effect of novel atmospheric-pressure jet pretreatment on the drying kinetics and quality of white grapes. *Journal of the Science of Food and Agriculture*, *99*(11), 5102–5111. https://doi.org/10.1002/JSFA.9754

Jain, D. (2007). Modeling the performance of the reversed absorber with packed bed thermal storage natural convection solar crop dryer. *Journal of Food Engineering, 78*(2), 637–647. https://doi.org/10.1016/J.JFOODENG.2005.10.035

Janjai, S., Intawee, P., Kaewkiew, J., Sritus, C., & Khamvongsa, V. (2011). A large-scale solar greenhouse dryer using polycarbonate cover: Modeling and testing in a tropical environment of Lao People's Democratic Republic. *Renewable Energy, 36*(3), 1053–1062. https://doi.org/10.1016/J.RENENE.2010.09.008

Janjai, S., Srisittipokakun, N., & Bala, B. K. (2008). Experimental and modelling performances of a roof-integrated solar drying system for drying herbs and spices. *Energy, 33*(1), 91–103. https://doi.org/10.1016/J.ENERGY.2007.08.009

Jin, W., Mujumdar, A. S., Zhang, M., & Shi, W. (2017). Novel drying techniques for spices and herbs: A Review. *Food Engineering Reviews, 10*(1), 34–45. https://doi.org/10.1007/S12393-017-9165-7

Kaewkiew, J., Nabnean, S., & Janjai, S. (2012). Experimental investigation of the performance of a large-scale greenhouse type solar dryer for drying chilli in Thailand. *Procedia Engineering, 32*, 433–439. https://doi.org/10.1016/J.PROENG.2012.01.1290

Kareem, M. W., Habib, K., Ruslan, M. H., & Saha, B. B. (2017). Thermal performance study of a multi-pass solar air heating collector system for drying of Roselle (*Hibiscus sabdariffa*). *Renewable Energy, 113*, 281–292. https://doi.org/10.1016/J.RENENE.2016.12.099

Khallaf, A. E. M., & El-Sebaii, A. (2022). Review on drying of the medicinal plants (herbs) using solar energy applications. *Heat and Mass Transfer, 58*(8), 1411–1428. https://doi.org/10.1007/S00231-022-03191-5

Kumar, R. N., Natarajan, M., & Natarajan, K. (2020). Analysis of solar tunnel dryer performance with red chili drying in two intervals. *Research Journal of Chemistry and Environment, 24*(1), 125–129.

Kurup, A. H., Deotale, S., Rawson, A., & Patras, A. (2020). Thermal processing of herbs and spices. *Herbs, Spices and Medicinal Plants*, 1–21. https://doi.org/10.1002/9781119036685.CH1

Lakshmi, D. V. N., Muthukumar, P., Ekka, J. P., Nayak, P. K., & Layek, A. (2019). Performance comparison of mixed mode and indirect mode parallel flow forced convection solar driers for drying *Curcuma zedoaria*. *Journal of Food Process Engineering, 42*(4), e13045. https://doi.org/10.1111/JFPE.13045

Łechtańska, J. M., Szadzińska, J., & Kowalski, S. J. (2015). Microwave- and infrared-assisted convective drying of green pepper: Quality and energy considerations. *Chemical Engineering and Processing: Process Intensification, 98*, 155–164. https://doi.org/10.1016/j.cep.2015.10.001

Łyczko, J., Jałoszyński, K., Surma, M., García-Garví, J. M., Carbonell-Barrachina, Á. A., & Szumny, A. (2019). Determination of various drying methods' impact on odour quality of true lavender (*Lavandula angustifolia* Mill.) Flowers. *Molecules, 24*(16). https://doi.org/10.3390/molecules24162900

Mandal, R., Singh, A., & Pratap Singh, A. (2018). Recent developments in cold plasma decontamination technology in the food industry. *Trends in Food Science & Technology, 80*, 93–103. https://doi.org/10.1016/J.TIFS.2018.07.014

Marnoto, T. (2014). Drying of Rosella (*Hibiscus sabdariffa*) flower petals using solar dryer with double glass cover collector. *International Journal of Science and Engineering, 7*(2), 150–154. https://doi.org/10.12777/ijse.7.2.150-154

Mihindukulasuriya, S. D. F., & Jayasuriya, H. P. W. (2015). Drying of chilli in a combined infrared and hot air rotary dryer. *Journal of Food Science and Technology, 52*(8), 4895–4904. https://doi.org/10.1007/s13197-014-1546-9

Miraei Ashtiani, S. H., Salarikia, A., & Golzarian, M. R. (2017). Analyzing drying characteristics and modeling of thin layers of peppermint leaves under hot-air and infrared treatments. *Information Processing in Agriculture, 4*(2), 128–139. https://doi.org/10.1016/j.inpa.2017.03.001

Nalawade, S. A., Ghiwari, G. K., & Hebbar, H. U. (2019). Process efficiency of electromagnetic radiation (EMR)-assisted hybrid drying in spearmint (*Mentha spicata* L.). *Journal of Food Processing and Preservation*, *43*(11), 1–10. https://doi.org/10.1111/jfpp.14190

Nayak, S., Kumar, A., Mishra, J., & Tiwari, G. N. (2011). Drying and testing of mint (*Mentha piperita*) by a hybrid photovoltaic-thermal (PVT)-based greenhouse dryer. Drying Technology, *29*(9), 1002–1009. https://doi.org/10.1080/07373937.2010.547265

Nimnuan, P., & Nabnean, S. (2020). Experimental and simulated investigations of the performance of the solar greenhouse dryer for drying cassumunar ginger (*Zingiber cassumunar* Roxb.). *Case Studies in Thermal Engineering*, *22*, 100745. https://doi.org/10.1016/J.CSITE.2020.100745

Nour-Eddine, B., Belkacem, Z., & Abdellah, K. (2013). Experimental study and simulation of a solar dryer for spearmint leaves (*Mentha spicata*). *International Journal of Ambient Energy*, *36*(2), 50–61. https://doi.org/10.1080/01430750.2013.820149

Nukulwar, M. R., & Tungikar, V. B. (2022). Recent development of the solar dryer integrated with thermal energy storage and auxiliary units. *Thermal Science and Engineering Progress*, *29*, 101192. https://doi.org/10.1016/J.TSEP.2021.101192

Pandey, A., & Pandey, A. (2017). Harvesting and post-harvest processing of medicinal plants: Problems and prospects. *The Pharma Innovation Journal*, *6*(12), 229–235. https://www.researchgate.net/publication/322076894

Pankaew, P., Aumporn, O., Janjai, S., Pattarapanitchai, S., Sangsan, M., & Bala, B. K. (2020). Performance of a large-scale greenhouse solar dryer integrated with phase change material thermal storage system for drying of chili. *International Journal of Green Energy*, *17*(11), 632–643. https://doi.org/10.1080/15435075.2020.1779074

Pei, Y., Li, Z., Song, C., Li, J., Song, F., Zhu, G., & Liu, M. (2020). Effects of combined infrared and hot-air drying on ginsenosides and sensory properties of ginseng root slices (*Panax ginseng* Meyer). *Journal of Food Processing and Preservation*, *44*(1), 1–13. https://doi.org/10.1111/jfpp.14312

Poós, T., & Varju, E. (2017). Drying characteristics of medicinal plants. *International Review of Applied Sciences and Engineering*, *8*(1), 83–91. https://doi.org/10.1556/1848.2017.8.1.12

Puente-Díaz, L., Ah-Hen, K., Vega-Gálvez, A., Lemus-Mondaca, R., & Scala, K. D. (2013). Combined infrared-convective drying of murta (*Ugni molinae* Turcz) berries: Kinetic modeling and quality assessment. *Drying Technology*, *31*(3), 329–338. https://doi.org/10.1080/07373937.2012.736113

Rabha, D. K., Muthukumar, P., & Somayaji, C. (2017). Experimental investigation of thin layer drying kinetics of ghost chilli pepper (Capsicum Chinense Jacq.) dried in a forced convection solar tunnel dryer. *Renewable Energy*, *105*, 583–589. https://doi.org/10.1016/j.renene.2016.12.091

Romero, V. M., Cerezo, E., Garcia, M. I., & Sanchez, M. H. (2014). Simulation and validation of vanilla drying process in an indirect solar dryer prototype using CFD fluent program. *Energy Procedia*, *57*, 1651–1658. https://doi.org/10.1016/J.EGYPRO.2014.10.156

Romuli, S., Schock, S., Nagle, M., Chege, C. G. K., & Müller, J. (2019). Technical performance of an inflatable solar dryer for drying amaranth leaves in Kenya. *Applied Sciences*, *9*(16), 3431. https://doi.org/10.3390/APP9163431

Saeidi, K., Jafari, S., Hosseinzadeh Samani, B., Lorigooini, Z., & Doodman, S. (2020). Effect of some novel and conventional drying methods on quantitative and qualitative characteristics of Hyssop essential oil. *Journal of Essential Oil-Bearing Plants*, *23*(1), 156–167. https://doi.org/10.1080/0972060X.2020.1723443

Saifullah, M., McCullum, R., McCluskey, A., & Vuong, Q. (2019). Effects of different drying methods on extractable phenolic compounds and antioxidant properties from lemon myrtle dried leaves. *Heliyon*, *5*(12), e03044. https://doi.org/10.1016/j.heliyon.2019.e03044

Selvi, K. Ç., Kabutey, A., Gürdil, G. A. K., Herak, D., Kurhan, Ş., & Klouček, P. (2020). The effect of infrared drying on color, projected area, drying time, and total phenolic content of rose (*Rose electron*) petals. *Plants*, *9*(2). https://doi.org/10.3390/plants9020236

Sharma, A., Chen, C. R., & Vu Lan, N. (2009). Solar-energy drying systems: A review. *Renewable and Sustainable Energy Reviews*, *13*(6–7), 1185–1210. https://doi.org/10.1016/J.RSER.2008.08.015

Sharma, M., Atheaya, D., & Kumar, A. (2021). Recent advancements of PCM based indirect type solar drying systems: A state of art. *Materials Today: Proceedings*, *47*, 5852–5855. https://doi.org/10.1016/J.MATPR.2021.04.280

Shreelavaniya, R., Kamaraj, S., Subramanian, S., Pangayarselvi, R., Murali, S., & Bharani, A. (2021). Experimental investigations on drying kinetics, modeling and quality analysis of small cardamom (*Elettaria cardamomum*) dried in solar-biomass hybrid dryer. *Solar Energy*, *227*, 635–644. https://doi.org/10.1016/J.SOLENER.2021.09.016

Taheri-Garavand, & Meda, V. (2018). Drying kinetics and modeling of savory leaves under different drying conditions. *International Food Research Journal*, *25*(4), 1357–1364.

Thamkaew, G., Sjöholm, I., & Galindo, F. G. (2020). A review of drying methods for improving the quality of dried herbs. *Critical Reviews in Food Science and Nutrition*, *61*(11), 1763–1786. https://doi.org/10.1080/10408398.2020.1765309

Tran, T. T. A., & Nguyen, H. V. H. (2018). Effects of spray-drying temperatures and carriers on physical and antioxidant properties of lemongrass leaf extract powder. *Beverages*, *4*(4). https://doi.org/10.3390/beverages4040084

Yadav, Y. K. (2017). Performance evaluation of solar tunnel dryer for drying of garlic. *Current Agriculture Research Journal*, *5*(2), 220. https://doi.org/10.12944/CARJ.5.2.09

Yuan, Y., Lin, W., Mao, X., Li, W., Yang, L., Wei, J., & Xiao, B. (2019). Performance analysis of heat pump dryer with unit-room in cold climate regions. *Energies*, *12*(16), 3125. https://doi.org/10.3390/EN12163125

4 Impact of Drying on Various Quality Characteristics of Herbs, Spices, and Medicinal Plants

Kar Yong Pin
Forest Research Institute Malaysia

CONTENTS

DOI: 10.1201/9781003269250-4

4.1   INTRODUCTION

Generally, spices are derived from various plant parts including bark, buds, flowers, fruits, leaves, rhizomes, roots, seeds, stigmas, and styles or the entire plant tops (UNIDO & FAO, 2005). Herbs are mainly referred to plants whose leaves are used as flavor in culinary. Besides, spices and herbs also function as preservatives and therapeutic agents. The unique applications of these plant products are mainly attributed to the presence of essential oils and oleoresins. However, these aromatic organic compounds are mostly heat-sensitive in nature.

Medicinal plants are classified as plants that possess therapeutic values which can be used for various medical treatments. The prescriptions of these plants are well-documented in the ancient scripts like Ayurveda and Traditional Chinese Medicine. Scientific studies have proven that the beneficial effects are contributed by the bioactive phytochemicals found in different plant parts. These phytochemicals provide balanced and synergic actions that are lacking in single-compound synthetic drugs. Examples of several popular medicinal plants with its bioactive phytochemicals and benefits are given in Table 4.1.

TABLE 4.1
Popular Medicinal Plants with Their Bioactive Phytochemical and Benefits

Medicinal Plants	Bioactive Phytochemicals	Benefits
Ginseng (*Panax ginseng*)	Ginosides	Enhance both the central and immune systems. Inhibit oxidative stress that causes certain chronic diseases and aging.
Holy basil (*Ocimum sanctum* L.)	Eugenol, rosmarinic acid, apigenin, myretenal, luteolin, β-sitosterol, and carnosic acid	Anti-diabetic, wound healing, antioxidant, radiation protective, immunomodulatory, anti-fertility, anti-inflammatory, anti-microbial, anti-stress, and anti-cancer activities.
Neem (*Azadirachta indica*)	Azadirachtin, nimbolinin, nimbin, nimbidin, nimbidol, salannin, and quercetin	Antioxidant, anti-cancer, anti-inflammatory, hepatoprotective, and anti-malarial.
Tongkat Ali (*Eurycoma longifolia* Jack)	Quassinoids including eurycomanone, eurycomanols, hydroxyklaineanones, eurycomalactones, eurycomadilactones, eurylactones, laurycolactones, longilactones, and hydroxyglaucarubol	Used as an aphrodisiac, antibiotic, appetite stimulant, and health supplement.
King of Bitters (*Andrographis paniculata*)	Diterpenoids including andrographolide, neoandrographolide, and dehydroandrographolide	Anti-bacterial, anti-inflammatory, anti-thrombotic, and hepatoprotective properties

Drying is one of the critical steps in post-harvest processing of spices, herbs, and medicinal plants, and it plays a decisive role in determining the dried product quality. The requirements in drying of these valuable plants are as follows: (i) moisture content is reduced to a level that inhibits microbial and chemical activities during storage; (ii) minimum quality degradation in terms of active ingredients, color, flavor, and aroma; and (iii) microbial count must be below the prescribed limit (Rocha et al., 2011).

This chapter aims to discuss the impacts of pretreatment and various drying methods, namely, freeze drying (FD), heat pump drying (HPD), microwave drying (MD), microwave vacuum drying (MVD), dual-stage drying-convective pre-drying coupled with microwave vacuum finish drying (CPD-MVFD), infrared drying (IRD), and radio-frequency drying (RFD) on the key quality characteristics of dried spices, herbs, and medicinal plant products.

4.2   IMPORTANT QUALITY CHARACTERISTICS

4.2.1   RESIDUAL MOISTURE CONTENT

Residual moisture content (RMC) is defined as the total amount of water present in a product after drying. The stability of a product is greatly influenced by RMC. High RMC leads to microbial growth and chemical reactions that degrade the product quality. The MC for spice, herbs, and medicinal plants is usually described in percentage following the equation below:

$$MC = \frac{M_w - M_d}{M_w} \qquad (4.1)$$

where MC is the moisture content, M_w is the initial product weight (g), and M_d is the final product weight (g) after drying.

As given in Table 4.2, the acceptable RMC is subjected to the drying method and product characteristic. The RMC of garlic and onion products is relatively low due to their hygroscopic nature. These products tend to absorb moisture from the surrounding during the storage and hence susceptible to microbial contamination. Low RMC is also required for most leafy products.

The permissible RMC of medicinal plants is prescribed in various pharmacopoeias developed by the respective countries, regional authorities, and international organizations. There is no correlation between RMC with plant parts used (Rocha et al., 2011). Typically, the RMC for dried medicinal plants is in the range of 8%–12%.

4.2.2   WATER ACTIVITY

Water activity is defined as the ratio of the vapor pressure of water in a material to the vapor pressure of pure water at the same temperature. The limits of water activity are imposed to prevent the growth of spoilage microorganisms. European Spice Association (ESA) recommends that the water activity of spices must be lower than 0.65. American Spice Trade Association (ASTA) establishes the limit of water activity depending on the particular microorganisms including bacteria, fungi, yeast, and

TABLE 4.2

Residual Moisture Content (RMC) of Selected Herbs and Spices (European Spice Association)

Product	Maximum RMC (%)
Garlic products	6.5
Onion products	6.0–8.0 (depending on origin)
Leaf	
Parsley leaves	7.5
Celery leaves	8.0
Coriander leaves	8.0
Laurel leaves	8.0
Lemon Grass	10.0
Oregano	12.0
Seed/Rhizome/Fruit	
Celery seed	11.0
Coriander seed	12.0
Cloves	12.0
Ginger	12.0
Pepper black	12.0
Pepper white	12.0
Pepper green	13.0 (8.0 if freeze-dried)
Pepper pink (Schinus)	14 (8.0 if freeze-dried)

molds as given in Table 4.3. Generally, water activity below 0.6 is sufficient to inhibit microbial proliferation. The achievable water activity is often limited by the drying method and material characteristics.

Water present in a material could facilitate chemical and enzymatic reactions during storage. These reactions alter phytochemical content and thus change the odor, flavor, and bioactivity of the materials. Lower water activity limits the "free" water that causes the breakdown of the beneficial chemicals and thus enhances the stability of the product during storage.

4.2.3 COLOR

Color is a quality parameter that affects consumer's desire in purchasing spices and herbs for culinary purposes. The dried spices and herbs are preferable to have color as close to its original color as possible. However, its importance in the medicinal plants sector is not as critical because consumers would focus more on its therapeutic properties. Most of the drying methods, except FD and other low-temperature drying methods, cause significant change of color in plant materials. The color degradation is due to the deterioration of plant pigments including chlorophylls, anthocyanins, carotenoids, and betalains. These naturally derived colorants are highly susceptible to pH, thermal stress, and light (Dikshit and Tallapragada, 2018).

TABLE 4.3

Minimum Water Activity for Growth of Spoilage Microorganisms Commonly Found in Spice by ASTA

Microorganism	Minimum Water Activity
Bacteria	
Salmonella	0.93–0.94
C. perfringens	0.97
C. botulinum	0.94
Bacillus cereus	0.93
E. coli	0.95
Listeria monocytogenes	0.92
Halobacterium halobius	0.75
Staphylococcus aureus	0.82
Yeast and mold	
Saccharomyces cerevisiae	0.90
Zygosaccharomyces rouxii	0.62
Aspergillus flavus/parasiticus	0.80
Botrytis cinerea	0.97
Penicillium ssp.	0.79–0.82
Rhizopus stolonifer	0.89
Xeromyces bisporus	0.61

4.2.4 PHYTOCHEMICALS

There are two types of metabolites produced in plants, namely, primary and secondary metabolite. Primary metabolites are essential for physiological processes while secondary metabolites are created in response to interactions with environment to mediate the ecological stress and establish the defense mechanism. Secondary metabolites provide protections against various natural hazards such as ultraviolet (UV) irradiation, herbivores, and pathogen attacks (Xiao et al., 2017b). Figure 4.1 illustrates the components of primary and secondary metabolites produced in plants.

The plant essences of interest are small organic molecules (mol. Wt. < 2,000 amu approx.), which are secondary metabolites with different functional groups and polarities. Secondary metabolites consist of three major groups, namely, terpenoids, phenolics, and alkaloids (Adeyemi, 2011; de Padua et al., 1999), which are critical for plant survival in neutralizing biotic and abiotic stress from the growing surroundings. These compounds also exhibit diverse therapeutic properties that are useful in promoting human's health. Nowadays, scientists and researchers call these plant-originated molecules collectively as phytochemicals.

These organic phytochemicals are produced by plant cells and stored in different parts including glandular trichomes, plant tissue, cell membrane, and vacuoles. Trichomes are uni- or multicellular appendages covering the plant surface and are produced by most of the plant species (Werker, 2000). These are generally classified into two types, namely, non-glandular and glandular. Phytochemicals are secreted

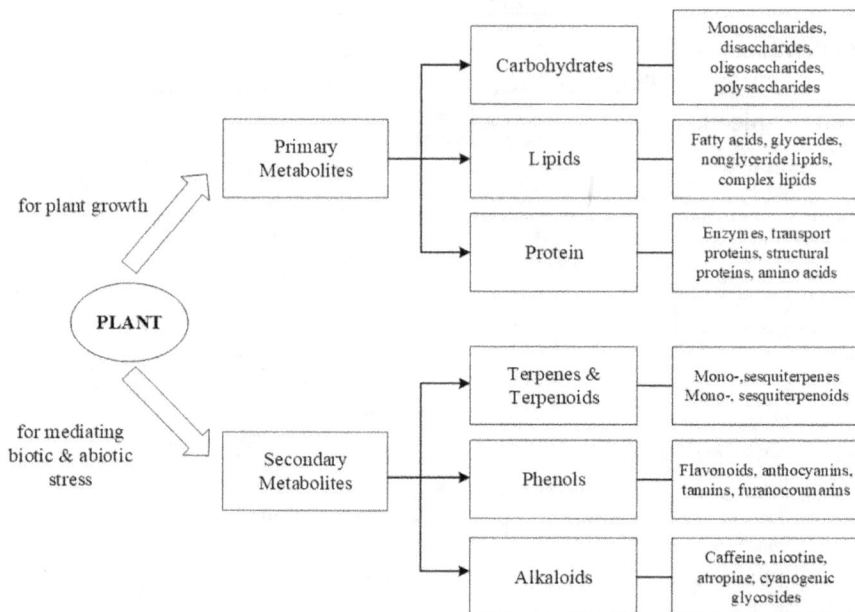

FIGURE 4.1　Primary and secondary metabolites produced by plant.

and stored in glandular trichomes for the defensive mechanism. Beckman et al. (1972) reported that phenols were found in the hair-like trichomes of 32 of 43 plant species in his study.

Trichomes on the leaf surface secrete and release secondary metabolites including alkanes, acyl sugars, and sesquiterpenes to repel or kill herbivorous insects (Xiao et al., 2017b). The flavonoids in the form of water-soluble glycosides are secreted by glandular trichomes of *Phillyrea latifolia* leaves when exposed to excessive light (Tattini et al., 2000). Bin and Qing (2007) reported that flavonoids can adsorb UV-B radiation and protect the epidermal plant tissues from radiation damage.

Drying process induces considerable damage to plant parts in which the phytochemicals are stored. The exposure of heat leads to loss or degradation of volatile and thermolabile compounds. It also triggers chemical reactions such as acid hydrolysis, enzymatic reaction, and oxidation reaction that break down complex phytochemicals and form new compounds. As prescribed in Ayurveda and Traditional Chinese Medicine, medicines are prepared using decoction of a mixture of dried plant materials. Rupture of cell structure caused by drying is favorable in the extraction of phytochemicals as the solvent is able to penetrate easily into the plant matrix and extract the active constituents.

4.2.5　Essential Oils

Essential oils are obtained from the leaves, flowers, stems, bark, seeds, fruits, roots, and plant exudates conventionally by steam distillation. It has a wide range of applications in medicine, food, cosmetic, and aromatherapy industries.

Nowadays, there are about 300 types of commercially available essential oils which are approximately 10% of the identified essential oils. An essential oil may contain up to several hundred volatile aromatic compounds, and this complex mixture of compounds gives the oil its distinct characteristic (UNIDO & FAO, 2005). Essential oils usually comprise two or three major components in relatively high concentrations (20%–95%), and other components are present in trace levels (Shaaban et al., 2012). The aromatic composition is influenced by the type of herb, age of the plant, harvesting season, post-harvest practices, and storage condition (Thamkaew et al., 2021).

Basically, essential oils consist of 90%–95% volatile fraction and 5%–10% non-volatile fraction (De Castro et al., 1999). Non-volatile fraction contains hydrocarbons, fatty acids, sterols, carotenoids, waxes, coumarins, psoralens, and flavonoids. Volatile fraction consists of terpenes and terpenoids, along with aliphatic aldehydes, alcohols, and esters.

Due to the presence of high percentage of volatile compounds, exposure to hot surroundings and drying often leads to a significant loss of essential oils. In the essential oil industry, fresh plants are processed within the shortest possible time after harvesting to obtain the maximum yield. Chauhan et al. (2016) reported that a loss of 6.52% and 10.87% of essential oil in *Cymbopogon distans* (Nees ex Steud.) Wats. was observed over a fresh herb after being stored in field conditions for 7 and 15 days, respectively. The effects of drying treatment on the recovery and composition of essential oils vary depending on the properties of the volatiles, the type of drying method, and process parameters.

4.2.6 ANTIOXIDANT PROPERTIES

Numerous oxidative stress-related diseases are associated with free radicals and reactive oxygen species that are generated either within the human body or external sources. As biological functions in human body decline due to aging, the inherent protective mechanism fails to maintain the balance between free radicals and antioxidants. These reactive species attack and damage the functional molecules including DNA, proteins, and lipids and thus give rise to the occurrence of severe disorders.

Spices, herbs, and medicinal plants are major sources of antioxidants. Secondary metabolites including phenols, tannin, and flavonoids contribute to the antioxidant activity of plant materials. These naturally occurring reactive oxygen scavengers neutralize the reactions of free radicals and protect the body from adverse effects.

Reports showed that drying process could yield various effects on antioxidant properties in which no change, increase, or decrease phenomena were observed. The decrease is mainly due to the thermal and enzymatic degradation of antioxidant-related compounds. On the other hand, the release of bound phenolic compounds by cell disruption and formation of new antioxidants following heat treatment resulted in the enhancement of antioxidant properties (Chan et al., 2013). Plant cells are still intact and active during the initial stage of drying. Abiotic stress caused by heat and moisture loss triggers the protective mechanisms in cells to produce more antioxidant phytochemicals.

4.3 PRETREATMENT BEFORE DRYING

Pretreatment before drying has been widely applied in the fruit and vegetable industry, but the studies of its application in herbs, spices, and medicinal plants are not comprehensive. The purposes of pretreatment include inactivating enzymes, enhancing drying rate and product quality, minimizing non-enzymatic browning reactions, removing pesticide residues and toxic constituents, expelling air in plant tissues, decreasing microbial load, and increasing the extraction efficiency of bioactive compound (Xiao et al., 2017a). Generally, pretreatment methods are categorized as thermal and non-thermal. The available thermal methods are hot water, steam, microwave, ohmic, and high-humidity hot air impingement. Meanwhile, the non-thermal methods include high pressure, ultrasound, pulse electric field, and osmotic dehydration.

Dokhani et al. (2012) studied the microwave pretreatment on the quality *Achillea filipendulina* flowers and leaves before MVD. The plant materials were pretreated using microwave at 1,800 W for 2.5 minutes. The untreated flowers contained higher total volatiles but lower total phenolic content as compared to the pretreated samples. There was no significant difference in total volatiles of pretreated and untreated leaves. However, the phenolic content was higher in the untreated leaves. High microwave radiation generated a substantial heat with the leaves. The long treatment time (150 seconds) under intense heat caused degradation of phenolic compounds.

Phoungchandang and Kongpim (2012) observed that sweet basil after hot water blanching preserved better than untreated leaves in terms of phenolics and antioxidant activity in two drying methods, namely, convective drying (CD) and HPD. The leaves were blanched in boiling water for 60 seconds before drying. The blanching pretreatment caused a structural damage in the plant matrix and facilitated the release of phytochemicals.

Trirattanapikul and Phoungchandang (2014) studied the effects of hot water blanching (HWB) and microwave pretreatment under different treatment times on *Centella asiatica* leaves. The indirect contact thermal method, microwave pretreatment, performed better than HWB in all the studied treatment times in retaining the phenolic compounds and antioxidant activity. It was concluded that microwave pretreatment at 800 W for 30 seconds was the best setting. The microwave-treated leaves also contained higher phenolic content and antioxidant activity than untreated leaves after being dried with CD and HPD.

In a study involving CD, HPD, MVD, and FD, the blanched *Andrographis paniculata* leaves showed greater retention of total phenolics, major chemical constituents, and antioxidant activity than untreated leaves except the FD samples (Tummanichanont et al., 2017). The highest quality in terms of the three evaluation parameters was found in the unblanched FD leaves. The HWB at 95°C–98°C for 15 seconds resulted in a considerable chemical degradation.

Klungboonkrong et al. (2018) compared the effects of HWB and vacuum blanching (VB) on the inactivation of peroxidase enzyme and content of major compounds in *Orthosiphon aristatus* leaves. The major compounds studied were sinensetin and eupatorin. The leaves were directly dipped into boiling water in

HWB, while the leaves were packed in vacuum bag before dipped into boiling water in VB. The blanching time required to inactivate the enzyme was 45 and 75 seconds for HWB and VB, respectively. The content of the major compounds was higher in VB as compared to HWB. The indirect contact during VB prolonged the time required for enzyme inactivation but prevented the loss of active phytochemicals which could diffuse into the blanching water. The results from the same study also showed that VB leaves contained higher phenolics, major compounds, and antioxidant activity as compared to unblanched samples in all drying methods including CD, HPD, and FD.

Steam blanching and microwave pretreatment in MD of parsley leaves were compared by evaluating its effects on drying time, energy consumption, lutein content, chlorophyll content, and color (Sledz et al., 2016). Both pretreatment methods showed positive effect in reducing the drying time and energy consumption in comparison to untreated leaves, but there was no conclusive difference between the two methods. Both methods also did not show significant influence in lutein content, and there was no clear correlation of chlorophyll retention with respect to the type of pretreatment. The color of the parsley leaves obtained after steam blanching and drying at 40°C/300 W and after US treatment and dehydration at 30°C/300 W were very close to the fresh leaves.

Pulse electric field pretreatment was reported to accelerate the drying process of basil (*Ocimum basilicum* l.) leaves and its intensity greatly affected the product quality (Kwao et al., 2016). The treated leaves, with conditions that stomata were irreversibly opened while the rest of the tissue remains viable, were superior in terms of aroma compounds and color and have better rehydration capacity than the untreated control. Irreversible tissue damage provoked by high intensity (65 pulses of 1,500 V/cm, 150 µs pulse width, 760 µs between pulses) compromised the quality characteristics of the dried leaves as compared to other dried products.

Bozkir et al. (2018) reported that ultrasound pretreatment increased the drying rate by 19.30% and 13.82%, respectively, in hot air drying and MD of garlic slices. The ultrasound pretreatment could create cavitation and channels within the sample that facilitated the transfer of moisture during drying. Besides that, the cavitation also helped in breaking the bonds between the water molecules and thus increased the drying rate.

Osae et al. (2019) reported that the drying time of dried ginger (*Zingiber officinale* Roscoe) slices was reduced by 50.0%, 44.4%, and 33.3% for osmosonication (OS), ultrasound (US), and osmotic dehydration (OD) pretreatment, respectively, as compared to the untreated slices. Microstructure analysis showed that the pretreatments caused the deformation of cell structure and resulted in creation of inter-cellular space and micro-channels. The combined effects of osmotic pressure and ultrasound in OS pretreatment led to the most significant damage in cell structure and micro-channel formation which accelerated the subsequent drying process.

The study on the pretreatment for drying of herbs, spices, and medicinal plants is still relatively scarce. Generally, pretreatment is advantageous in shortening the drying time due to the formation of micro-channels for effective mass transfer. The quality of the pretreated plant materials is also better, and the selection of pretreatment methods and time are critical in determining its product quality.

4.4   DRYING METHODS

4.4.1   Freeze Drying

FD is basically a vacuum drying method that operates below the triple point of water (0.01°C and 0.61 kPa). Water in the food product forms ice during the freezing step, and the subsequent moisture removal is achieved by sublimation of ice crystals with well-controlled temperature and pressure. The ice evaporates from the products during vacuum-drying steps and leaves behind minute voids. The porous structure allows the product to maintain its original shape and results in crispy texture as well as rapid rehydration.

This method is superior in preserving the quality of food products as compared to other drying methods. FD is often used as the quality benchmark for comparison in drying studies. It can preserve most of the physical, chemical, and biological properties of the products because of its unique operating conditions including low drying temperature, absence of oxygen, and minimum level of free water within the product. The product also remains in the frozen state throughout the whole drying duration. However, FD is a slow process because heat can only be delivered from the heating plate to the material by conduction. Its shortcomings are high capital and operating costs because FD runs under high vacuum and requires refrigeration.

Tummanichanont et al. (2017) reported that the content of total phenolics, two major constituents, and antioxidant activity of freeze-dried *Andrographis paniculata* leaves was higher than convective-dried leaves. FD was also reported to retain higher phytochemicals and antioxidant properties than hot air in drying of black locus flowers, torch ginger, mulberry, blue trumpet vine, and lemon myrtle leaves (Ji et al., 2012; Chan et al., 2013; Saifullah et al., 2019).

The freeze-dried herbs including bay leaf, parsley, and basil contained lesser essential oils as compared to convective-dried samples (Díaz-Maroto et al., 2002a, 2002b; Díaz-Maroto et al., 2004). Convective-dried thyme under 40°C produced 4.6% more total volatiles than freeze-dried samples (Calin-Sanchez et al., 2013). Nofer et al. (2018) reported that the content of total volatiles in cepe mushroom dried at 70°C and 80°C was about 62.3% higher than FD. In a similar study, recovery of volatile compounds in black locust flowers dried with CD (50°C, 60°C and 70°C) was about 6.9% in average higher than FD (Ali et al., 2020).

Plant cells produce volatile terpenoids and reduce their volatility with glyco-sylation and oxidation. The enzymatic activities produce glycosylated form of the volatile compounds. These compounds are more stable for accumulation and storage because of the increase in hydrophilicity, molecular weight, and formation of inter-molecular hydrogen bonds (Yazaki et al., 2017). Heat treatment during drying induces the reverse reactions of glycosylated terpenoids which resulted in higher yield in CD as compared to low-temperature FD.

Freeze-dried dill managed to retain 58% of total volatile compound while only 11% in dill that was convective-dried at 50°C (Huopalahti et al., 1985). FD was reported as a better method in retaining the characteristic volatile compounds of basil compared to convective and MD (Cesare et al., 2003). Antal et al. (2011) reported a higher retention (43.1%) of essential oils in freeze-dried spearmint leaves as compared to convective-dried samples. FD was also found to be more effective in preserving the major volatiles in flower and leaf of *Achillea filipendulina* Lam (Dokhani et al., 2012). Morshedloo et al. (2021) concluded that FD was more suitable than CD in dehydration of aerial parts of Iranian dragonhead allowing high-essential oil yields.

The retention of volatile compounds is closely related to the chamber pressure during FD. The content of total volatiles in freeze-dried spearmint leaves at high pressure (150–250 Pa for 14 hours) was about 18.5% higher than those at low pressure (10–30 Pa for 12 hours) (Antal et al., 2011). Antal et al. (2014) reported freeze-dried lemon balm leaves at low pressure (50–80 Pa for 12 hours) suffered 34.8% loss of total volatiles as compared to 14.0% loss in samples dried using high pressure (250–300 Pa for 14 hours). Glandular trichomes in the leaves dried with low pressure showed greater structural damage which caused more volatile oils evaporated (Antal et al., 2014).

4.4.2   HEAT PUMP DRYING

The major feature of HPD is dehumidification of drying air developed based on the thermodynamic cycle used in the refrigeration method (Kudra and Mujumdar, 2009). The dehumidification unit is coupled with a module to heat up the inlet air to the desired temperature. Many studies showed that HPD is more energy efficient than some conventional drying methods such as CD and solar drying.

In a drying study of Kaprao leaves (white and red holy basil), Phoungchandang et al. (2003) concluded that HPD was able to retain eugenol and methyl eugenol higher than CD. Phoungchandang and Kongpim (2012) concluded that the quality of sweet basil, in terms of antioxidant activity and total phenolic, from HPD was better than those convective-dried materials under the same drying temperatures (40°C, 50°C and 60°C). In a similar study, *Centella asiatica* leaves dried using HPD showed higher total phenolics content and antioxidant activity than those from CD (Trirattanapikul and Phoungchandang, 2014).

HPD was also used in the drying of dill (*Anethum graveolens*) greens in comparison to hot air drying, infrared-hot air drying, and RFD (Naidu et al., 2016). The content of chlorophyll, carotenoids, and ascorbic acid HPD samples was comparable to the fresh drill. HPD (50°C ± 2°C and 28%–30% RH) required a longer drying time (270 minutes) as compared to infrared-hot air drying (240 minutes) and RFD (210 minutes) but faster than hot air drying (50°C ± 2°C, 58%–63% RH).

Tummanichanont et al. (2017) reported that total phenolics, major compounds (andrographolide and neoandrographolide), and antioxidant activity of *Andrographis paniculata* leaves from HPD were slightly higher than CD but lower than FD. The minor difference was because there was no significant difference in the drying time for HPD and CD run at the same temperatures (40°C, 50°C and 60°C). Low-temperature drying process in FD minimized the degradation of phytochemicals and thus produced leaves with better quality.

In a study of heat pump-assisted solar drying, Gan et al. (2017) found that there were no significant differences between the fresh and dried samples in the retention of sinensetin and rosmarinic acid. The samples from solar drying system without heat pump showed much lower retention of the two major compounds, and the significant loss could be due to a longer drying time (4 days) was required.

In all the studies, the quality of the plant materials, including total phenolics, antioxidant activity, and the targeted marker compounds, dropped under an increase in drying temperature in HPD, but the quality was still better than those dried using CD because of a shorter drying time in HPD.

4.4.3 MICROWAVE DRYING

Microwave is defined as electromagnetic radiations in the frequency range of 300 MHz–300 GHz and induces rapid heating by transmitting its energy via ionic conduction and dipole rotation of polar molecules. The electromagnetic energy is dissipated evenly on the surface and into the core of the material, and thus leads to a rapid and uniform drying.

The quality of products from MD is affected by microwave power, drying time, initial moisture content, and properties of the products (Moses et al., 2014). Dielectric permittivity and loss factor are the two properties that affect the efficiency of MD. Dielectric permittivity is the material ability in absorbing the electromagnetic energy, while loss factor is defined as its ability to transfer the energy into heat (Hayat et al., 2019).

The phenolic content in microwave-dried peppermint leaves (700 W) was higher than convective-dried (50°C) and fresh samples (Arslan et al., 2010). In the drying of fennel seed, Hayat et al. (2019) reported that both microwave and CD increased the total phenolic content and total flavonoids content, but MD was more efficient as it produced better results with shorter drying time. The findings also showed that prolonged heating time led to quality reduction.

Paramanandam et al. (2021) concluded that microwave-dried figs contained higher amount of phenolics, tannins, and flavonoids than figs dried from CD and sun drying. This is mainly due to the difference in drying time in which about 72 hours for sun drying and CD while only 10 minutes duration in MD. High microwave power is better in retaining the antioxidant activity because of shorter exposure to heat. In a drying study of *Andrographis paniculata* leaves, MVD performed better than CD in preserving phenolic compounds, two major phytochemicals, and antioxidant activity but not as good as FD (Tummanichanont et al., 2017). It was also found that these important properties were in a decline with the increase in microwave power (270, 450 and 720 W).

A greater loss of the total volatiles content within 44%–88% in Marjoram dried with MD (175, 385 and 595 W) was observed as compared to convective-dried samples (45°C ± 2°C) which only suffered 7% in loss (Raghavan et al., 1997). The authors pointed out that the loss of volatiles was closely proportional to the increase in microwave power, but high microwave power was favorable to color preservation due to the shorter drying time. Rao et al. (1998) reported similar findings in drying of rosemary in which greater loss of volatiles in microwave-dried samples (175, 385 and 595W) was observed as compared to convective-dried (45°C) and fresh samples.

In a study of ginger drying, Kubra and Rao (2012) reported that the loss of total volatiles in convective-dried samples (10.3%) was higher than samples that were dried under 800 W microwave power (6.6%). However, its loss was lower when compared to those using 385, 505, and 600 W with an average loss of 32%. MD with 800 W also contained about 47% more zingiberene (the major compound of ginger oil) than CD (50°C ± 4°C) and slightly more than the fresh ginger. Huang et al. (2012) also reported similar findings in which the content of zingiberene in microwave-dried samples (700 W) was higher than convective dried (80°C) and fresh ginger.

Incorporation of microwave in drying addresses surface overheating in the CD especially in material with low conductivity. The volumetric heating by electromagnetic energy increases the heat transfer and thus shortens the drying time that preserves better product quality.

4.4.4 MICROWAVE VACUUM DRYING

Vacuum is applied in drying to reduce the drying temperature and escalate the rate of moisture removal. The oxidation of phytochemicals is also minimized with the absence of oxygen under vacuum. The major setback of this method is heat transfer to the material only by conduction. Microwave is widely incorporated in vacuum drying to provide direct heating to high-polar water molecules within the whole volume of the material. This volumetric heating effect leads to rapid moisture removal and prevent structural collapse. This process creates a porous structure and reduces product density (Sham et al., 2001), and this puffing phenomenon is similar to FD.

Vacuum level and microwave power are the two key parameters in determining the product quality. Properties of the targeted phytochemicals in response to microwave are also affecting the drying results. The loss of volatile compounds and other bioactive phytochemicals caused by MVD is largely dependent on the polarity. In general, the condition of choice in MVD is high vacuum and low microwave power. The drying process is slow in this condition but still shorter than CD and FD (Szumny et al., 2010).

Scanning electron micrographs showed that the microstructure of mint leaves dried with MVD was more porous and open as compared to convective-dried leaves (Therdthai and Zhou, 2009). Disruption of cell structure and extracellular matrix takes place as water molecules absorb microwave energy to reach the super-heated state during the drying process. These structural alterations facilitate the release of phytochemicals from plant cells and penetration of solvent into plant matrix and thus increase the yield of active compounds. On the other hand, these conditions could exaggerate the loss or degradation of phytochemicals especially under prolonged heat treatment.

In the drying of *Roninia pseudoacacia* L. flowers, the total phenolic content and antioxidant properties of flowers from MVD were comparable to FD (Ji et al., 2012). Ali et al. (2020) reported that the total phenolic content in Malaysian rosemary in MVD at 240 W (2,144.51 mg GAE/100g dw), 360 W (2,352.50 mg GAE/100g dw), and 480 W (1,989.26 mg GAE/100g dw) was significantly higher than FD (199.58 mg GAE/100g dw), but FD performed better in preserving the anti-diabetic and anti-aging properties. Comparisons of the effects of MVD, FD, and CD on total phenolic content in different studies are summarized in Table 4.4.

TABLE 4.4
Preservation of Total Phenolics Using MVD, FD, and CD

Plant	Total Phenolics Content	Test Method	Operating Conditions	Reference
Flowers (*Robinia pseudoacacia* L.)	FD > MVD > CD	Folin-Cicocalteau phenol method (gallic acid equivalent)	CD: 60°C, 240 min MVD: 1,500 W, 70 kPa, 40 min FD: 12 hours	Ji et al. (2012)
Flowers (*Achillea filipendulina* L.)	CD > FD > MVD	Glories' method (Chlorogenic acid equivalent)	CD: 65°C, 420 minutes MVD: 450–1,800 W, 4.0 kPa, 12 minutes FD: 5 days	Dokhani et al. (2012)
Leaves (*Achillea filipendulina* L.)	MVD > FD > CD			
Leaves (Pink Rock Rose)	MVD ≅ CD	Folin-Cicocalteau phenol method (Gallic acid equivalent)	CD: 60°C, 240 minutes MVD: 240 W, 4–6 kPa, 480 minutes	Stepien et al. (2019)
Leaves (Pink Rock Rose)	CD > MVD > FD	UPLC-PDA-Q/TPF-MS method	CD: 60°C, 240 minutes MVD: 240 W, 4–6 kPa, 480 minutes FD: n.a	Matłok *et al.* (2020)
Leaves (*Phyla nodiflora*)	CD > MVD	Folin-Cicocalteau phenol method (Gallic acid equivalent)	CD: 60°C, 120 minutes MVD: 240 W, 4–6 kPa, 21 minutes	Chua et al. (2019b)
Leaves (*Strobilanthes crispus*)	CD > MVD > FD	Folin-Cicocalteau phenol method (Gallic acid equivalent)	CD: 60°C, 120 minutes MVD: 360 W, 4–6 kPa, 21 minutes	Chua et al. (2019c)
Leaves (Rosemary)	CD > MVD > FD	Folin-Cicocalteau phenol method (Gallic acid equivalent)	CD: 70°C, 120 minutes MVD: 360 W, 4–6 kPa, <40 minutes FD: n.a	Ali et al. (2020)

In the drying of rosemary, the highest total volatile recovered after MVD operated at 0 kPa and 240 W was 68.3% with reference to the fresh samples, while the recovery from CD was 64.6% (Calin-Sanchez et al., 2011). Figure 4.2 shows that under constant microwave power of 360 W, the concentration of total volatiles in rosemary dropped when higher vacuum level was used (e.g. 96–98 kPa). Hence, the higher the vacuum level under constant microwave power the higher the power intensity tends to reduce the concentration of total volatiles.

The effect of microwave power on recovery of total volatile compounds depends on the response of the volatiles present in plants to microwave energy (Table 4.5). Generally, high microwave power tends to reduce substantially the total volatiles in the dried materials.

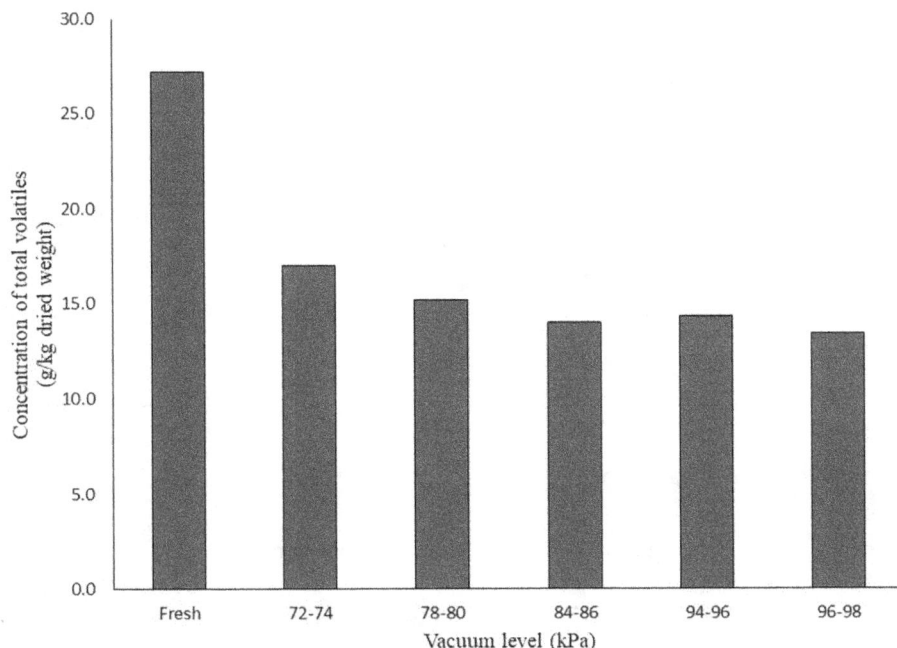

FIGURE 4.2 Concentration of total volatiles in rosemary dried under different vacuum level using constant microwave power (360 W).

TABLE 4.5
Recovery of Total Volatiles in Herbs at Different Microwave Power in MVD

Plant	Recovery of Total Volatile Compounds (%)			Reference
	240 W	360 W	480 W	
Rosemary[a]	45.6	52.6	42.6	Calin-Sanchez et al. (2011)
Basil	44.6	-	42.1	Calin-Sanchez et al. (2012)
Thyme	118.2	104.9	83.0	Calin-Sanchez et al. (2013)
Lavender	34.9	28.6	42.3	Łyczko et al. (2019)
Leaves (*Phyla nodiflora*)	9.1	27.2	16.9	Chua et al. (2019b)
Leaves (*Strobilanthes crispus*)	2.7	2.2	1.1	Chua et al. (2019c)
Rosemary (Malaysia)	28.3	28.2	28.1	Ali et al. (2020)

[a] The vacuum level was 94–96 kPa only for this study and the rest were 4–6 kPa.

The recovery of total volatiles of oregano from MVD was 84.5% as compared to only 30.9% from CD (Figiel et al., 2010). Łyczko et al. (2019) reported MVD was more effective than CD in preserving the total volatiles compounds in lavender.

Interestingly, analysis of the content of volatile compounds indicated that linalyl acetate increased in CD and MVD with 240 W. Linalyl acetate is one of the main compounds that contribute to the lavender fragrance. The increase of this compound after CD was two times the increase in MVD. This observation could be due to the formation of linalyl acetate by the plant during the direct-contact thermal stress induced by CD.

Comparisons on total volatile compounds in various herbs dried with MVD, FD, and CD in recent studies were summarized in Table 4.6. The comparisons are made based on the best results from each drying method.

TABLE 4.6
Comparisons of Total Volatiles Content Using MVD, FD, and CD

Plant	Total Volatile Compounds	Operating Conditions	Reference
Leaves (Oregano)	MVD > CD	CD: 60°C, 130 minutes MVD: 360 W, 4–6 kPa, 24 minutes	Figiel et al. (2010)
Flowers (*Achillea filipendulina* L.)	MVD > FD > CD	CD: 65°C, 7 hours MVD: 450–1,800 W, 4.0 kPa, 12 minutes	Dokhani et al. (2012)
Leaves (*Achillea filipendulina* L.)	FD > MVD > CD	FD: 5 days	
Leaves (Thymes)	MVD > FD, CD FD ≅ CD	CD: 40°C, 1,585 minutes MVD: 240 W, 4–6 kPa, 28 minutes FD: 24 hours	Calin-Sanchez et al. (2013)
Leaves (Lavender)	MVD > CD	CD: 70°C, 135 minutes MVD: 480 W, 4–6 kPa, 14 minutes	Łyczko et al. (2019)
Leaves (Basil)	MVD ≅ CD	CD: 40°, 1,744 minutes MVD: 240 W, 4–6 kPa, 28 minutes	Calin-Sanchez et al. (2012)
Leaves (Rosemary)	CD > MVD	CD: 60°C, 3 hours MVD: 480 W, 4–6 kPa, 30 minutes	Szumny et al. (2010)
Cepe mushroom	CD > FD > MVD	CD: 80°C, 210 minutes MVD: 240 W, 4–6 kPa, 72 minutes FD: 24 hours	Nofer et al. (2018)
Leaves (*Phyla nodiflora*)	CD > MVD	CD: 60°C, 120 minutes MVD: 240 W, 4–6 kPa, 21 minutes	Chua et al. (2019b)
Leaves (*Strobilanthes crispus*)	CD > FD > MVD	CD: 50°C, 150 minutes MVD: 240 W, 4–6 kPa, 28 minutes FD: not given	Chua et al. (2019c)
Leaves (Rosemary)	CD > MVD > FD	CD: 60°C, < 330 minutes MVD: 360 W, 4–6 kPa, <40 minutes FD: n.a	Ali et al. (2020)
Leaves (Pink Rock Rose)	CD > FD > MVD	CD: 60°C, 240 minutes MVD: 240 W, 4–6 kPa, 480 minutes FD: n.a	Matłok et al. (2020)

4.4.5 DUAL-STAGE DRYING – CONVECTIVE PRE-DRYING COUPLED WITH MICROWAVE VACUUM FINISH DRYING

MVD is relatively high in operating cost because high energy input is required for both applications of microwave and vacuum during drying. Researchers have investigated the practicality of combining CD and MVD to reduce the usage of MVD especially in the early drying stage. CD is effective in the removal of surface moisture in the initial stage, and MVD is then applied at the later stage when the drying kinetics begins to lose its linear characteristics in the rate of moisture removal (Chua et al., 2019a). The dual-stage drying process is named Convective Pre-Drying – Microwave Vacuum Finish Drying (CPD-MVFD).

Stepien et al. (2019) reported that there was no significant difference in the total phenolic content and antioxidant activity of Pink Rock Rose leaves dried using CPD-MVFD (50°C and 240 W) and MVD (240 W). However, the total phenolic content in leaves from CPD-MVFD was about 10% higher than CD at 60°C. In the drying of *Strobilanthes crispus* leaves, CPD-MVFD (50°C and 360 W) performed poorly in retaining phenolic compounds and antioxidant activity as compared to CD (40°C) and MVD with a microwave power of 360 W (Chua et al., 2019c). Chua et al. (2019b) also reported that the total phenolic content in *Phyla nodiflora* leaves from CPD-MVFD (50°C and 360 W) was significantly higher than those from CD (50°C) and MVD (240 W), but the antioxidant activity in CPD-MVFD was much lower than those from CD and MVD.

Ali et al. (2020) conducted a comprehensive study on the effects of CD, MVD, FD, and CPD-MVFD on Malaysian rosemary. However, no clear trend could be concluded on the effects on antioxidant, anti-diabetic, and anti-aging activity. In general, FD showed the best performance in preserving the bioactivities, but samples from FD contained the lowest total phenolics. The total phenolic content of CPD-MVFD samples was significantly lower than those from CD and MVD.

The effects of CPD-MVFD on the total volatile content of different plant materials in comparison to other drying methods are summarized in Table 4.7. Overall, CPD-MVFD is a dual-stage drying process that could improve the efficiency of CD and MVD.

4.4.6 INFRARED DRYING

Infrared drying (IRD) utilizes the heat that is generated by electromagnetic energy from infrared. It is only suitable for thin-layer drying because of the limitation in infrared penetration power. This drying method provides several advantages including adaptability, simplicity, fast heating rate, and high drying rate (Thamkaew et al., 2021).

Changes in volatiles, chemical profile, antioxidant properties, and microstructure of Chinese ginger (*Zingiber officinale* Roscoe) dried using air drying, FD, infrared drying, MD, and intermittent microwave – convective drying (IMCD) were investigated to determine the optimum drying method (An et al., 2016). The results indicated that air drying and infrared drying performed well in preserving the volatile oils. Ginger from FD, infrared drying, and IMCD were reported showing higher total

TABLE 4.7

Comparisons of Total Volatiles Content Using CPD-MVFD, MVD, FD, and CD

Plant	Total Volatile Compounds	Operating Conditions	Reference
Leaves (Rosemary)	CPD-MVFD > CD > MVD	CD: 60°C, 180 minutes MVD: 480 W, 4–6 kPa, 30 minutes CPD-MVFD: 60°C, 360 W, 88 minutes	Szumny et al. (2010)
Leaves (Basil)	CPD-MVFD > MVD, CD	CD: 40°C, 1,744 minutes MVD: 240 W, 4–6 kPa, 28 minutes CPD-MVFD: 40°C, 360 W, 252 minutes	Calin-Sanchez et al. (2012)
Leaves (*Phyla nodiflora*)	CPD-MVFD > CD > MVD	CD: 60°C, 120 minutes MVD: 240 W, 4–6 kPa, 21 minutes CPD-MVFD: 50°C, 360 W, 105 minutes	Chua et al. (2019b)
Leaves (Pink Rock Rose)	CPD-MVFD, CD > FD > MVD	CD: 60°C, 240 minutes MVD: 240 W, 4–6 kPa, 480 minutes FD: n.a	Matłok *et al.* (2020)
Leaves (Oregano)	MVD > CD, CPD-MVFD	CD: 60°C, 130 minutes MVD: 360 W, 4–6 kPa, 24 minutes CPD-MVFD: 60°C, 360 W, 93 minutes	Figiel et al. (2010)
Leaves (Thymes)	MVD > CPD-MVFD > FD, CD FD ≅ CD	CD: 40°C, 1,585 minutes MVD: 240 W, 4–6 kPa, 28 minutes FD: 24 hours CPD-MVFD: 40°C, 240 W, 301 minutes	Calin-Sanchez et al. (2013)
Leaves (Lavender)	MVD > CPD-MVFD, CD	CD: 70°C, 135 minutes MVD: 480 W, 4–6 kPa, 14 minutes CPD-MVFD: 60°C, 480 W, 64 minutes	Łyczko et al. (2019)
Leaves (*Strobilanthes crispus*)	CD > FD > MVD, CPD-MVFD	CD: 50°C, 150 minutes MVD: 240 W, 4–6 kPa, 28 minutes FD: n.a CPD-MVFD: 50°C, 360 W, 105 minutes	Chua et al. (2019c)
Leaves (Rosemary)	CD > CPD-MVFD > MVD > FD	CD: 60°C, < 330 minutes MVD: 360 W, 4–6 kPa, <40 minutes FD: n.a CPD-MVFD: 50°C, 360 W, 30 minutes	Ali et al. (2020)

phenolic content, total flavonoids content, and antioxidant activities. However, FD and infrared drying were higher in energy consumption.

Sellami et al. (2012) reported that the highest essential oil yields from sage (*Salvia officinalis* L.) plants were obtained by infrared drying at 45°C (0.39%) followed by air drying (0.30%) and oven drying at 45°C (0.26%). A significant loss of about 99.9% was observed when the drying temperature of IRD and oven drying was increased to 65°C.

Ning and Han (2013) investigated the application of far-infrared drying of taegeuk ginseng (*Panax ginseng* C.A. Meyer) in comparison to hot air drying. The hot air drying time was reported taking 3–3.5 days longer than far-infrared drying under similar conditions. The energy consumption in far-infrared was about 9.67%–14.8% less than that required by hot air drying. The quality of taegeuk ginseng by far-infrared drying in terms of color, saponin, and polyphenol contents was also superior as compared to hot air drying.

Nguyen et al. (2015) compared the performance of six drying methods in the drying of *Phyllanthus amarus,* namely, hot air, low-temperature air, infrared, microwave, sun, and vacuum drying. The results indicated that MD was the fastest among the six methods. Infrared drying at 30°C was found to retain the highest bioactive compound yield as well as maximum antioxidant capacity.

A hybrid infrared-hot air drying performed better than hot air drying in terms of drying rate and retention of aroma quality of star anise (*Illicium Verum*) (Wen et al., 2020). The drying rate of infrared-hot air drying was higher than those from hot air drying in all the temperatures (60°C, 70°C and 80°C) and sun drying. The volatile oil content was the highest in star anise from sun drying in which the temperature range was 21.5°C–35.5°C and followed by infrared-hot air drying. Low temperature and rapid drying process minimized the loss of volatile oil. Besides that, infrared and ultraviolet radiation induced damage to plant cells and seized reactions that degrade the volatile oil.

The reported works in the infrared drying of spices, herbs, and medicinal plants are limited but its potential in improving the product quality is indeed promising. The shortcomings in the penetration power and energy consumption could be addressed by coupling with other drying methods.

4.4.7 RADIO-FREQUENCY DRYING

Radio-frequency drying employs the energy transmitted from the electromagnetic radiation in the frequency range of 10–300 MHz, and the US Federal Communications Commission (FCC) only allows three frequencies, namely, at 13.56 ± 6.68 kHz, 27.12 ± 160.00 kHz, and 40.68 ± 20.00 kHz for food applications (Zhou and Wang, 2019). Similar to microwave, RF induces volumetric heating that addresses the low thermal conductivity in conventional methods. In comparison to microwave, RF with longer wavelength has a greater penetration depth and thus provides a more uniform heating.

Naidu et al. (2016) compared the effect of hot air drying, HPD, infrared-hot air drying, and RFD in the drying of dill green. The drying curves showed that RFD with its effective dielectric heating required the shortest time among the methods investigated, but the samples from RFD suffered the greatest loss in chlorophyll, carotenoids, and ascorbic acid content.

Li et al. (2021) recommended that RFD has to be used as a potential drying method for dandelion (*Taraxacum mongolicum*) leaves before extracting the leaves for functional food and medicinal applications. It was observed that RFD produced dried leaves that showed high antioxidant activities, in-vitro anti-glycation activity, and in-vitro inhibitory activity on α-glucosidase as compared to other methods such as MVD, vacuum drying, FD, and hot air drying.

The effect of hot air-assisted radio-frequency drying (HARF) was compared with hot air drying on the quality of jujube (Song et al., 2022). The results showed that HARF required shorter time than hot air drying at 360 and 1,320 minutes, respectively. Jujube dried from HARF also exhibited better quality in terms of color, vitamin C, and total flavonoid.

Due to the limited literature on RFD, its effects in preserving product quality of spices, herbs, and medicinal plants are still inconclusive. The method is proven in improving the drying rate and reducing the energy consumption in food processing application. The application of RFD could also minimize microbial contamination because RF heating is widely used for food sterilization.

4.5   CONCLUSION

Drying is one of the important steps in post-harvest processing with the aim to prolong the shelf life of spices, herbs, and medicinal plants. It seizes the growth of bio-contaminants and unwanted chemical reactions that degrade the materials by removing its water content. Most of the drying methods involve heat treatment and might cause certain level of degradation of bioactive compounds. The loss of the phytochemicals would compromise the desired quality and bioactivity. Thus, one of the major considerations in drying is to preserve the active phytochemicals. Pretreatment before drying is an area that needs to be further explored because previous studies indicate that the additional step serves well in preserving or improving the product quality. Based on reported studies, it can be concluded that the selection of the optimum drying method and operating conditions is subjected to the prioritized quality characteristics of each plant. There is no universal principle that can be referred in addressing the complex reactions within plant materials that alter the product quality during drying.

REFERENCES

Adeyemi, M.M.H. (2011). A review of secondary metabolites from plant materials for post-harvest storage. *International Journal Pure and Applied Sciences and Technology*, 6(2), 94–102.

Ali, A., Oona, C.C., Chua, B.L., Figiel, A., Chong, C.H., Wojdylod, A., Turkiewiczd, I.P., Szumny, A., & Łyczkoe, J. (2020). Volatile and polyphenol composition, anti-oxidant, anti-diabetic and antiaging properties, and drying kinetics as affected by convective and hybrid vacuum microwave drying of *Rosmarinus officinalis* L. *Industrial Crops & Products*, 151, 112463.

An, K., Zhao, D., Wang, Z., Wu, J., Xu, Y., & Xiao, G. (2016). Comparison of different drying methods on Chinese ginger (*Zingiber officinale* Roscoe): Changes in volatiles, chemical profile, antioxidant properties, and microstructure. *Food Chemistry*, 197, 1292–1300.

Antal, T., Chong, C.H., Law, C.L., & Sikolya, L. (2014). Effects of freeze drying on retention of essential oils, changes in glandular trichomes of lemon balm leaves. *International Food Research Journal*, 21(1), 387–394.

Antal, T., Figiel, A., Kerekes, B., & Sikolya, L. (2011). Effect of drying methods on the quality of the essential oil of spearmint leaves (*Mentha spicata* L.). *Drying Technology*, 29(15), 1836–1844.

Arslan, D., Ozcan, M.M., & Menges, H.O. (2010). Evaluation of drying methods with respect to drying parameters, some nutritional and colour characteristics of peppermint (*Mentha x piperita* L.). *Energy Conversion and Management*, *51*(12), 2769–2775.

Beckman, C.H., Mueller, W.C., & McHardy, W.E. (1972). The localization of stored phenols in plant hairs. *Physiological Plant Pathology*, *2*(1), 69–74.

Bin, L., & Qing, Z. (2007). Effect of enhanced UV-B radiation on plant flavonoids. *Chinese Journal of Eco Agriculture*, *15*(3), 191–194.

Bozkir, H., Rayman Ergün A., Tekgül Y., & Baysal, T. (2018). Ultrasound as pretreatment for drying garlic slices in microwave and convective dryer. *Food Science and Biotechnology*, *28*(2), 347–354.

Calin-Sanchez, A., Figiel, A., Lech, K., Szumny, A., & Carbonell-Barrachina, A. A. (2012). Volatile composition of sweet basil essential oil (*Ocimum basilicum* L.) as affected by drying method. *Food Research International*, *48*(1), 217–225.

Calin-Sanchez, A., Figiel, A., Lech, K., Szumny, A., & Carbonell-Barrachina, A.A. (2013). Effects of drying methods on the composition of thyme (*Thymus vulgaris* L.) essential oil. *Drying Technology*, *31*(2), 224–235.

Calin-Sanchez, A., Szumny, A., Figiel, A., K. Jałoszynski, K., Adamski, M., & Carbonell-Barrachina, A.A. (2011). Effects of vacuum level and microwave power on rosemary volatile composition during vacuum–microwave drying. *Journal of Food Engineering*, *103*(2), 219–227.

Cesare, L.F.D., Forni, E., Viscardi, D., & Nani, R.C. (2003). Changes in the chemical composition of basil caused by different drying procedures. *Journal of Agricultural and Food Chemistry*, *51*(12), 3575–3581.

Chan, E.W.C., Lye, P.Y., Eng, S.Y., & Tan, Y.P. (2013). Antioxidant properties of herbs with enhancement effects of drying treatments: A synopsis. *Free Radicals and Antioxidants*, *3*(1), 2–6.

Chauhan, A., Verma, R.S., Padalia, R.C., Upadhyay, R.K., & Bahl, J.R. (2016). Post harvest storage effect on essential oil content and composition of *Cymbopogon distans* (Nees ex Steud.) Wats. *Journal of Essential Oil Research*, *28*(6), 540–544.

Chua, L.Y.W., Chong, C.H., Chua, B.L., & Figiel, A. (2019a). Influence of drying methods on the antibacterial, antioxidant and essential oil volatile composition of herbs: A review. *Food and Bioprocess Technology*, *12*, 450–476.

Chua, L.Y.W., Chua, B.L., Figiel, A., Chong, C.H., Wojdyło, A., Szumny, A., & Łyczko, J. (2019b). Drying of phyla nodiflora leaves: Antioxidant activity, volatile and phytosterol content, energy consumption, and quality studies. *Processes*, *7*(4), 210.

Chua, L.Y.W., Chua, B.L., Figiel, A., Chong, C.H., Wojdyło, A., Szumny, A., & Choong, T.S.Y. (2019c). Antioxidant activity, and volatile and phytosterol contents of *Strobilanthes crispus* dehydrated using conventional and vacuum microwave drying methods. *Molecules 24*(7), 1397.

De Castro, M.L., Jiménez-Carmona, M.M., & Fernandez-Perez, V. (1999). Towards more rational techniques for the isolation of valuable essential oils from plants. *TrAC Trends in Analytical Chemistry*, *18*(11), 708–716.

De Padua, L.S., Bunyapraphatsara, N., & Lemmens, R.H.M.J. (1999). *Plant Resources of South-East Asia*. Leiden: Backhuys Publishers.

Díaz-Maroto, M., Pérez-Coello, M., & Cabezudo, M. (2002a). Effect of different drying methods on the volatile components of parsley (*Petroselinum crispum* L.). *European Food Research and Technology*, *215*(3), 227–230.

Díaz-Maroto, M.C., Pérez-Coello, M.S., & Cabezudo, M.D. (2002b). Effect of drying method on the volatiles in bay leaf (*Laurus nobilis* L.). *Journal of Agricultural and Food Chemistry*, *50*(16), 4520–4524.

Díaz-Maroto, M.C., Sánchez Palomo, E., Castro, L., González Viñas, M.A., & Pérez-Coello, M.S. (2004). Changes produced in the aroma compounds and structural integrity of basil (*Ocimum basilicum* L) during drying. *Journal of the Science of Food and Agriculture*, *84*(15), 2070–2076.

Dikshit, R., & Tallapragada, P. (2018). Comparative study of natural and artificial flavoring agents and dyes. In A.M. Grumezescu, & A.M. Holban, (Ed.). *Natural and Artificial Flavoring Agents and Food Dyes* (pp. 83–111). New York: Academic Press.

Dokhani, S., Durance, T.D., Tony Cottrell, T., & Mazza, G. (2012). Drying effects on major volatile and phenolic components of *Achillea filipendulina* Lam. *Journal of Essential Oil Bearing Plants*, *15*(6), 885–894.

Figiel, A., Szumny, A., Gutiérrez-Ortíz, A., Carbonell-Barrachina, A.A. (2010). Composition of oregano essential oil (Origanum vulgare) as affected by drying method. *Journal of Food Engineering*, *98*(2), 240–247, https://doi.org/10.1016/j.jfoodeng.2010.01.002.

Gan, S.H., Tham, T.C., Ng, M.X., Chua, L.S., Aziz, R., Baba, M.R., Abdullah, L.C., Ong, S.P., & Law, C.L. (2017). Study on retention of metabolites composition in misai kucing (*Orthosiphon stamineus*) by heat pump assisted solar drying. *Journal of Food Processing and Preservation*, *41*(6), e13262.

Hayat, K., Abbas, S., Hussain, S., Shahzad, S.A., & Tahir, M.U. (2019). Effect of microwave and conventional oven heating on phenolic constituents, fatty acids, minerals and antioxidant potential of fennel seed. *Industrial Crops and Products*, *140*, 111610.

Huang, B., Wang, G., Chu, Z., & Qin, L. (2012). Effect of oven drying, microwave drying, and silica gel drying methods on the volatile components of ginger (*Zingiber officinale* Roscoe) by HS-SPME-GC-MS. *Drying Technology*, *30*(3), 248–255.

Huopalahti, R., Kesalahti, E., & Linko, R. (1985). Effect of hot air and freeze drying on the volatile compounds of dill (*Anethum graveolens* L.) herb. *Agricultural and Food Science, 57*(2), 133–138.

Ji, H., Du, A., Zhang, L., Xu, C., Yang, M., & Li, F. (2012). Effects of drying methods on antioxidant properties in *Robinia pseudoacacia* L. flowers. *Journal of Medicinal Plants Research*, *6*(16), 3233–3239.

Klungboonkrong, V., Phoungchandang, S., & Lamsal, B. (2018). Drying of *Orthosiphon aristatus* leaves: Mathematical modeling, drying characteristics, and quality aspects. *Chemical Engineering Communications, 205*(9), 1239–1251.

Kubra, I.R., & Rao, L.J.M. (2012). Effect of microwave drying on the phytochemical composition of volatiles of ginger. *International Journal of Food Science and Technology*, *47*(1), 53–60.

Kudra, T., & Mujumdar, A.S. (2009). *Advanced Drying Technologies*, 2nd ed., Boca Raton, FL: CRC Press.

Kwao, S., Al-Hamimi, S., Damas, M.E.V., Rasmusson, A.G., & Gomez Galindo, F. (2016). Effect of guard cells electroporation on drying kinetics and aroma compounds of genovese basil (*Ocimum basilicum* l.) leaves. *Innovative Food Science & Emerging Technologies*, *38*(A), 15–23.

Li, F., Feng, K.L., Yang, J.C., He, Y.S., Guo, H., Wang, S.P., Gan, R.Y., & Wu, D.T. (2021). Polysaccharides from dandelion (*Taraxacum mongolicum*) leaves: Insights into innovative drying techniques on their structural characteristics and biological activities. *International Journal of Biological Macromolecules*, *167*, 995–1005.

Łyczko, J., Jałoszyński, K., Surma, M., Masztalerz, K., Szumny, A. (2019). HS-SPME analysis of true lavender (*Lavandula angustifolia* Mill.) leaves treated by various drying methods. *Molecules*, *24*, 764. https://doi.org/10.3390/molecules24244538

Matłok, N., Lachowicz, S., Gorzelany, J., & Balawejder, M. (2020). Influence of drying method on some bioactive compounds and the composition of volatile components in dried pink rock rose (*Cistus creticus* L.). *Molecules*, *25*(11), 2596.

Morshedloo, M.R., Machiani, M.A., Mohammadi, A., Maggi, F., Aghdam, M.S., Mumivand, H., & Javanmard, A. (2021). Comparison of drying methods for the extraction of essential oil from dragonhead (*Dracocephalum moldavica* L., Lamiaceae). *Journal of Essential Oil Research, 33*(2), 162–170.

Moses, J.A., Norton, T., Alagusundaram, K., & Tiwari, B.K. (2014). Novel drying techniques for the food industry. *Food Engineering Reviews, 6*(3), 43–55.

Naidu, M.M., Vedashree, M., Satapathy, P., Khanum, H., Ramsamy, R., & Hebbar, H.U. (2016). Effect of drying methods on the quality characteristics of dill (*Anethum graveolens*) greens. *Food Chemistry, 192*, 849–856.

Nguyen, V.T., Vuong, Q.V., Bowyer, M.C., Van Altena, I.A., & Scarlett, C.J. (2015). Effects of different drying methods on bioactive compound yield and antioxidant capacity of *Phyllanthus amarus*, *Drying Technology, 33*(8), 1006–1017.

Ning, X., & Han, C. (2013). Drying characteristics and quality of taegeuk ginseng (*Panax ginseng* CA Meyer) using far-infrared rays. *International Journal of Food Science & Technology, 48*(3), 477–483.

Nofer, J., Lech, K., Figiel, A., Szumny, A., & Carbonell-Barrachina, A.A. (2018). The influence of drying method on volatile composition and sensory profile of *Boletus edulis*. *Journal of Food Quality*, 2018, 2158482.

Osae, R., Zhou, C., Xu, B., Tchabo, W., Bonah, E., Alenyorege, E.A., & Ma, H. (2019). Nonthermal pretreatments enhances drying kinetics and quality properties of dried ginger (*Zingiber officinale* Roscoe) slices. *Journal of Food Process Engineering, 42*(5), e13117.

Paramanandam, V., Jagadeesan, G., Muniyandi, K., Manoharan, A.L., Nataraj, G., Sathyanarayanan, S., & Thangaraj, P. (2021). Comparative and variability analysis of different drying methods on phytochemical, antioxidant and phenolic contents of *Ficus auriculata* Lour Fruit. *Phytomedicine Plus, 1*(3), 100075.

Phoungchandang, S., & Kongpim, P. (2012). Modeling using a new thin-layer drying model and drying characteristics of sweet basil (*Ocimum baslicum* Linn.) using tray and heat pump-assisted dehumidified drying. *Journal of Food Process Engineering, 35*(6), 851–862.

Phoungchandang, S., Sanchai, P., & Chanchotikul, K. (2003). The development of dehumidifiying dryer for a Thai herb drying (Kaprao leaves). *Food Journal, 33*, 145–155.

Raghavan, B., Rao, L.J., Singh, M., & Abraham, K.O. (1997). Effect of drying methods on the flavour quality of marjoram (*Oreganum majorana* L.). *Nahrung, 41*(3), 159–161.

Rao, L.J., Singh, M., Raghavan, B., & Abraham, K.O. (1998). Rosemary (*Rosmarinus officinalis* l.): Impact of drying on its flavor quality. *Journal of Food Quality, 21*(2), 107–115.

Rocha, R.P., Melo, E.C., & Radünz, L.L. (2011). Influence of drying process on the quality of medicinal plants: A review. *Journal of Medicinal Plants Research, 5*(33), 7076–7084.

Saifullah, M., McCullum, R., McCluskey, A., & Vuong, Q. (2019). Effects of different drying methods on extractable phenolic compounds and antioxidant properties from lemon myrtle dried leaves. *Heliyon, 5*(12), e03044.

Sellami, I.H., Rebey, I.B., Sriti, J., Rahali F.Z., & Marzouk, B. (2012). Drying sage (*Salvia officinalis* l.) plants and its effects on content, chemical composition, and radical scavenging activity of the essential oil. *Food Bioprocess Technology, 5*, 2978–2989.

Shaaban, H.A.E., El-Ghorab, A.H., & Shibamoto, T. (2012). Bioactivity of essential oils and their volatile aroma components: Review. *Journal of Essential Oil Research, 24*(2), 203–212.

Sham, P.W.Y., Scaman, C.H., & Durance, T.D. (2001). Texture of vacuum microwave dehydrated apple chips as affected by calcium pretreatment, vacuum level, and apple variety. *Journal of Food Science, 66*(9), 1341–1347.

Sledz, M., Wiktor, A., Rybak, K., Nowacka, M., & Witrowa-Rajchert D. (2016). The impact of ultrasound and steam blanching pre-treatments on the drying kinetics, energy consumption and selected properties of parsley leaves. *Applied Acoustics, 103*, Part B, 148–156.

Song, S., Huang, X., Liu, Y., & Zhang, Q. (2022). Drying characteristics and quality analysis of hot air-assisted radio frequency and hot-air drying of jujube (*Zizyphus jujube* Miller cv. Jinsixiaozao). *Engenharia Agrícola, 42*(1), e20210112

Stepien, A.E., Gorzelany, J., Matłok, N., Lech, K., & Figiel, A. (2019). The effect of drying methods on the energy consumption, bioactive potential and colour of dried leaves of Pink Rock Rose (*Cistus creticus*). *Journal of Food Science and Technology, 56*(5), 2386–2394.

Szumny, A., Figiel, A., Gutierrez-Ortiz, A., & Carbonell-Barrachina, A.A. (2010). Composition of rosemary essential oil (*Rosmarinus officinalis*) as affected by drying method. *Journal of Food Engineering, 97*(2), 253–60.

Tattini, M., Gravano, E., Pinelli, P., Mulinacci, N. & Romani, A. (2000). Flavonoids accumulate in leaves and glandular trichomes of *Phillyrea latifolia* exposed to excess solar radiation. *New Phytologist, 148*(1), 69–77.

Thamkaew, G., Sjöholm, I., & Galindo, F.G. (2021). A review of drying methods for improving the quality of dried herbs. *Critical Reviews in Food Science and Nutrition, 61*(11), 1763–1786.

Therdthai, N., & Zhou, W. (2009). Characterization of microwave vacuum drying and hot air drying of mint leaves (*Mentha cordifolia* Opiz ex Fresen). *Journal of Food Engineering, 91*(3), 482–489.

Trirattanapikul, W., & Phoungchandang, S. (2014). Microwave blanching and drying characteristics of *Centella asiatica* (L.) urban leaves using tray and heat pump-assisted dehumidified drying. *Journal of Food Science and Technology, 51*(12), 3623–3634.

Tummanichanont, C., Phoungchandang, S., & Srzednicki, G. (2017). Effects of pretreatment and drying methods on drying characteristics and quality attributes of *Andrographis paniculata*. *Journal of Food Processing and Preservation, 41*(6), e13310.

UNIDO & FAO. (2005). Herbs, spices and essential oils, post-harvest operations in developing countries. Retrieved from http://www.fao.org/3/a-ad420e.pdf.

Wen, Y., Chen, L., Li, B., Ruan, Z., & Pan, Q. (2020). Effect of infrared radiation-hot air (IR-HA) drying on kinetics and quality changes of star anise (*Illicium verum*). *Drying Technology, 39*(1), 90–103.

Werker, E. (2000). Trichome diversity and development. In *Advances in Botanical Research*, (vol. 31, pp. 1–35). New York: Academic Press. https://www.sciencedirect.com/science/article/pii/S0065229600310059

Xiao, H., Pan, Z., Deng, L., El-Mashad, H.M., Yang, X., Mujumdar, A.S., Gao, Z., & Qian Zhang, Q. (2017a). Recent developments and trends in thermal blanching – A comprehensive review. *Information Processing in Agriculture, 4*(2), 101–127.

Xiao, K., Mao, X., Lin, Y., Xu, H, Zhu, Y., Cai, Q., Xie, H., & Zhang, J. (2017b). Trichome, a functional diversity phenotype in plant. *Molecular Biology, 6*(1), 1000183.

Yazaki, K., Arimura, G., & Ohnishi, T. (2017). 'Hidden' terpenoids in plants: Their biosynthesis, localization and ecological roles. *Plant and Cell Physiology, 58*(10), 1615–1621.

Zhou, X., & Wang, S. (2019). Recent developments in radio frequency drying of food and agricultural products: A review. *Drying Technology, 37*(3), 271–286.

5 Instrumentation and Analyses of Bioactive Compounds and Ingredients

Eric Wei Chiang Chan
UCSI University

Joash Ban Lee Tan
Monash University Malaysia

Choon Hui Tan
UCSI University

CONTENTS

DOI: 10.1201/9781003269250-5

5.1 INTRODUCTION

Food processing treatments including drying operations could change the properties of food but also alter their bioactive properties (Chan et al., 2019). It is widely believed that drying and processing, especially with treatment that involves heat, would degrade bioactive compounds and reduce the value of functional foods. However, this is not always the case as the heat applied is often not enough to degrade compounds. For instance, pyrolysis of sugars only starts at temperatures above 160°C.

Such beliefs are common as studies on the effects of heat treatment are still somewhat lacking. Rather than reducing, heat may increase the bioactive properties of foods with the release of additional compounds (Chan et al., 2009). For instance, increases in antioxidants after drying have also been reported in more recent studies by Ng et al. (2020b) and Chan et al. (2019). Reduction in bioactive compounds such as antioxidants would more likely be due to leaching and degradation by enzymes such as polyphenol oxidase (Chan et al., 2015; Siow et al., 2022).

Instrumentation and analyses of bioactive compounds are thus crucial for determining the actual effect of drying on medicinal plants and functional foods. Furthermore, to ensure the efficacy of purported bioactive properties, study designs must take steps to identify invalid metabolic panaceas (IMPs) and pan-assay interference compounds (PAINs). IMPs and PAINs are fairly recent concepts being first discussed by Bisson et al. (2016).

All PAINs are IMPs but not all IMPs are PAINs. IMPs are a wider group of compounds, some of which may bind to specific drug targets, making them not PAINs (Bisson et al., 2016). For instance, hydroxycitric acid from *Garcinia cambogia* is one such example of an IMP, which is not a PAIN (Batsis et al., 2021). Hydroxycitric acid inhibits ATP citrate lyase, which prevents de novo lipogenesis and reduces obesity in rats. In humans, de novo lipogenesis inhibition is not a good weight management strategy and may cause lipotoxicity and metabolic stress (Solinas et al., 2015). IMPs and PAINs are often described as black holes in drug discovery. Computational modelling approaches and studies using cell and animal models may reduce the number of IMPs and PAINs.

5.2 COMMON CLASSES OF NATURAL PRODUCTS

Spices, herbs, and medicinal plants contain a plethora of biologically active compounds, many of which have been extensively studied throughout the years. Given the sheer diversity of compound classes, an overview of some of the major groups that are of both bioactivity research and general consumer interests is presented in this chapter.

5.2.1 Phenols

Phenolic compounds are usually crystalline and absorb UV and, as such, can be analysed using photodiode arrays (Das and Gezici, 2018). They are ubiquitous compounds present in all plants as products from the shikimate pathway, taking various forms ranging from simple phenols such as phenol or catechol to large polymeric tannins. The antioxidant, anti-inflammatory, and even anticancer activities of many spices, herbs, and medicinal plants are often attributed to phenolic compounds and polyphenols (a wide array of organic compounds with multiple phenolic units). Additionally, given their role in pathogen defence, phenols are also often attributed to antimicrobial and insecticidal properties.

Given that the only requirement to be a "phenol" is technically the presence of a phenolic unit, there are numerous classes of phenols. Of these, the most well-studied type of phenol is flavonoids, with their recognisable three-ringed skeleton (a product of the aforementioned shikimate pathway and acetate pathway), comprising two six-carbon phenyl rings (ring A and B) with an oxygen-containing six-membered heterocycle between them (ring C). These can exist as glycosides or aglycones and can be subdivided into various subclasses, including anthocyanins, flavanols, flavones, flavanones, flavonols, flavanonols and isoflavones (Ferreira et al., 2017). Despite their ubiquity in plants, flavonoids cannot be synthesised by humans and animals, making them important components in their diet.

Besides flavonoids, there are hydroxycinnamates, which are hydroxy derivatives of phenylpropanoids, with their eponymous C_6C_3 skeleton. This includes compounds such as caffeic acid, ferulic acid and chlorogenic acid; all of which are widely reported as bioactive and present in various plants (Vuolo et al., 2019). In a similar vein, hydroxybenzoic acids (C_6C_1) such as gallic acid and ellagic acid have been reported to exhibit a wide range of bioactivities, including antioxidant, anti-inflammatory, anticancer and antimutagenic (Soong and Barlow, 2006).

There are also tannins, which are likewise ubiquitous, and both hydrolysable tannins (e.g. geraniin) (Cheng et al., 2020) and non-hydrolysable tannins are of considerable interest in bioactivity research (Vuolo et al., 2019).

5.2.2 Terpenoids

Terpenoids (or isoprenoids) are physically oils or amorphous solids that do not absorb UV, and as such, alternative forms of detection are required (Das and Gezici, 2018). They are the largest class of plant-based natural products with over 40,000 structures derived from isoprene ($CH_2=C(CH_3)-CH=CH_2$). While many prominent aromatic compounds in essential oils are terpenoids, essential oils are also comprised of many other non-terpenoid compounds. Additionally, terpenoids also take other forms; common examples include carotenoids, cannabinoids and natural latex. The volatile nature and distinct aroma of many smaller terpenoids (particularly monoterpenes) have resulted in their use in a wide variety of products and applications.

Common examples include menthol (mint), limonene (citrus), citral (citrus, lemon myrtle, and lemongrass) and geraniol (rose), which have been used in various foods, cosmetics, perfumes, and even insecticides and pharmaceuticals. Many of these bioactive terpenoids have been widely studied, and still a considerable amount of research are ongoing until this very day. Examples of recent in vitro studies include antioxidant and neuroprotective activities of olive oil by-products (Montenegro et al., 2021), cytotoxic activity of *Premna serratifolia* Linn (Biradi and Hullatti, 2017) and nitric oxide inhibition of *Euonymus verrucosus* var. *pauciflorus* (Yang et al., 2019).

5.2.3 ALKALOIDS

Alkaloids are crystalline but their defining amine group does not lend itself well to UV detection (Das and Gezici, 2018). Many of the most potent and well-known plant-based bioactive compounds come from this class including cocaine, nicotine and morphine. The term "alkaloid" broadly describes organic compounds (often cyclic) with one or more nitrogen atoms in them, making an extremely diverse group of over 12,000 compounds in around 20% of plants (Ramawat et al., 2009). Recent studies on alkaloids have continued on the trend of demonstrating their bioactivity, ranging from the in vitro and in vivo anticancer activity observed in papaverine from *P. somniferum* (Inada et al., 2019); to the cardioprotective activity of rutaecarpine from *Evodia rutaecarpa* ('Wu Zhu Yu') both in vitro and in vivo (Tian et al., 2019); in vitro anti-chikungunya activity by tomatidine often found in the leaves of tomatoes (*Solanum lycopersicum*) (Troost et al., 2020); in vitro acetylcholinesterase inhibition of a novel alkaloid from *Portulaca oleracea* L. (common purslane) (Xiu et al., 2019); antibacterial activity exhibited by nigritanine from *Strychnos nigritana* against *Staphylococcus aureus* (Casciaro et al., 2019).

5.2.4 VITAMINS

Vitamins are often not considered natural products as they are primary metabolites required for growth, development and reproduction. They are structurally diverse as they are not chemically and biologically related but are instead classified because of their role as primary metabolites, i.e. micronutrients. As such, analytical techniques for vitamins are diverse but well established. For instance, vitamin C (ascorbic acid) can be measured using titration and high-performance liquid chromatography (HPLC) Ng and Kuppusamy (2019). Titration may require more training for an accurate analysis, but HPLC-based methods are less accurate due to potential overlapping with other compounds in the plant matrix.

Despite not being a secondary metabolite, vitamins are of interest in food processing as they can be obtained from the diet and can be reduced by processing. While it is true that heat can degrade vitamins, losses from heat are not as drastic as one may expect. Martinsen et al. (2020) reported that cooking at 85°C reduced the ascorbic acid content of strawberry jam by 20% but not raspberry jam. Ng and Kuppusamy (2019) even reported that heating increased the ascorbic acid content of bitter melon (*Momordica charantia*). Losses from cooking are mainly from leaching into the cooking media as observed by Lešková et al. (2006).

5.3 IMPs AND PAINs IN DRUG DISCOVERY

Drug discovery favours high-throughput screening enabled by combinatorial chemistry (Hajduk and Greer, 2007). Combinatorial chemistry is a process of randomly combining small molecules in the hopes of identifying drug leads from a large number of randomly generated molecules. The main limitation of this random shotgun approach is that its hit rate is disappointingly low. The role of bioprospecting natural products from medicinal plants and functional foods is to identify new pharmaco-phores to be fed into combinatorial chemistry libraries. This makes the process less random as pharmacophores can bind specifically to drug targets.

IMPs and PAINs drastically reduce the efficiency of high-throughput screening (Bajorath, 2021). IMPs are active molecules in vitro but are impossible to translate into activity in vivo. PAINs are a similar group of compounds that bind non-specifi-cally to proteins and other molecules, often leading to a false-positive by aggregation. Therefore, it is important for the analytical study design to identify pharmacophores within molecules and subject these pharmacophores to molecular docking studies before the more expensive drug trials.

5.4   GENERAL APPROACHES

Natural products from plants have played a key role in drug discovery, but this has since expanded to include animals, bacteria and fungi (Atanasov et al., 2021). The first step involves screening for a biologically active extract followed by consecutive bioactivity-guided isolation. Although this is a time-consuming process, it is neces-sary to link a compound with a specific bioactivity. The different steps involved in bioassay-guided fractionation are tabulated in Table 5.1.

TABLE 5.1
Bioactivity-Guided Isolation

Step	Problem	Solution	Reference
Extraction	Different solvents yield different natural products	Optimise extraction solvents after identifying active compounds	Atanasov et al. (2021)
Screening	Natural products may vary according to the variety and growing conditions	Collect samples from different locations	Chan et al. (2007), Wong et al. (2009), and Siow et al. (2022)
Column chromatography	Similar compounds may be difficult to resolve	Use column packing material with different selectivity	Tan et al. (2015) and Lim et al. (2021)
Structural elucidation	Samples may be too little for structural elucidation	Start with a larger amount of plant samples of at least 1 kg and use a large initial column	Tan et al. (2015) and Lim et al. (2021)
Verify activity	Compounds may lose their activity after purification and some compounds may be IMPs and PAINs	The interaction of compounds in a solution and with target molecules can be studied using computational chemistry. This would also filter out many IMPs and PAINs	Bisson et al. (2016), Bazzo et al. (2020), and Magalhães et al. (2022)

5.4.1 Extraction and Screening

The most common extraction method is percolation in a solvent (Ng et al., 2020a). The polarity index is the ability of the solvent to interact with various polar test solutes as defined by Snyder (1974) and it is a widely used indicator for solvents in chromatography. Different compounds are extracted based on the polarity index and it can be adjusted to improve yields of the desired compound or exclude impurities. Methanol is often used because it has a middling polarity index of 5.1, ranging from 10.2 for water to 0.1 for hexane.

The middling polarity index of methanol ensures that it can extract a wide range of polar and non-polar natural products (Tan et al., 2015). Coupled with its low boiling point, methanol is the ideal solvent for screening because it extracts a wide range of compounds. However, this is also a major limitation; the wide range of compounds extracted means that there are also more compounds to be separated in the extract.

Using a more polar solvent like water or an aqueous mixture of water and methanol could be a good alternative if the target compounds are very polar such as heavily glycosylated flavonoids (Tan et al., 2020). However, the main limitation of water as an extraction solvent is that it is susceptible to bacterial growth which could compromise the sample's integrity. Mixing water with at least 50% v/v of methanol prevents bacterial growth, but most compounds extracted by aqueous solvents cannot diffuse past the cell membrane without active transport. This is highlighted by Lipinski's rule (Atanasov et al., 2021).

A hydrophobic solvent like hexane or solvent with a reduced polarity index such as ethyl acetate may be desirable to exclude very polar compounds (Lim et al., 2021). Hydrophobic compounds are of interest because they can diffuse past cell membranes. However, solubilising hydrophobic compounds in aqueous solutions for bioassays would be a challenge, and they may not have enough hydrogen bond acceptors and donors to interact with the target proteins properly. This is also a concern as highlighted by Lipinski's rule (Atanasov et al., 2021).

Once a suitable solvent has been identified, screening is a simple matter of testing the extract using desired bioassays (Tan et al., 2015; Lim et al., 2021) and some common bioassays are described in Section 5.5. Bioactivities are often expressed as an equivalent value relative to a standard against the extract weight or plant weight. Common examples include gallic acid equivalent per 100 g of fresh plant material, gallic acid equivalent per gram of dried plant material or gallic acid equivalent per gram of extract. Effective concentration 50 (EC_{50}) and inhibitory concentration 50 (IC_{50}) are also viable ways of expressing bioactivity. However, expressing values as a direct % bioactivity or concentration of standard is not recommended as bioactivity is often dose-dependent and it is difficult to compare these units with reported literature values.

With screening, natural product content can be similar regardless of where the plant is grown. For example, tea planted in Cameron Highlands has the same phenolic content as those planted in the coastal town of Banting in Malaysia, and hibiscus leaves have mostly the same phenolic content planted in the mountains or along the coast (Chan et al., 2007; Wong et al., 2009). The variations in tea flavour are primarily due to how the leaves are processed, i.e. drying and fermentation (Chan et al., 2007).

Growing conditions can alter the natural product content, but variations are more often seen in cultivated plants like cocoa that have been subjected to selective breeding. Siow et al. (2022) reported different flavour profiles and antioxidant content in single-origin cocoa beans from Malaysia, Vietnam and Venezuela. Differences could be due to the variety of cocoa beans from Malaysia and Vietnam of Trinitario variety and those from Venezuela of Criollo variety. However, differences were observed even between the Malaysia and Vietnam Trinitario variety cocoa beans possibly due to different climatic and soil conditions.

5.4.2 COLUMN CHROMATOGRAPHY AND STRUCTURAL ELUCIDATION

Column chromatography is probably the most time-consuming part of bioactivity-guided isolation. The distinction between preparative chromatography and analytical chromatography described in Section 5.6 is an important one (Schmidt-Traub et al., 2020). The former prioritises sample load for adequate yield for bioactivity testing, and the reproducibility of analyte retention time is not a key consideration. The latter prioritises good resolution of peaks for quantification and reproducible retention times for matching with standards. Prioritising resolution in preparative chromatography is a common mistake.

The purity and yield of compounds are important considerations for structural elucidation. Structural elucidation requires the use of nuclear magnetic resonance spectroscopy (NMR) in cases where mass spectrometry (MS) databases are not available. NMR works on the premise that atomic nuclei with an odd mass or charge have the same magnetic moment and are affected equally by radiofrequency pulses.

Any nuclei with an odd mass or charge can be analysed, but the two most relevant ones for natural products are ^{13}C and ^{1}H NMR. Functional groups will have different degrees of shielding, which would reduce the effects of radiofrequency pulses on the nuclear spin to give a measurable chemical shift compared to the standard which is usually tetramethylsilane for ^{13}C and ^{1}H NMR. Different variations of ^{13}C and ^{1}H NMR are described in Table 5.2.

Ten milligrams of the sample with 90% purity is often recommended for a good NMR spectrum, although proficient users can do less. Purification is achieved by a repeated round of chromatography. A column packing material with different selectivity is selected after each round of chromatography to help separate unresolved bands.

Table 5.3 shows a list of commonly used packing material and their selectivity. Most packing materials have a primary mechanism of selectivity and secondary mechanisms of selectivity. Partition chromatographic packing uses a liquid stationary phase and mainly retains hydrophobic compounds. Adsorption chromatography uses a solid stationary phase which, in addition to polarity, also differentiates compounds according to their shape. Reverse phase adsorption is often a macroporous styrene–divinylbenzene resin which also has the secondary selectivity of π-bond interactions. The π-bond interaction strongly interacts with aromatic groups, strongly retaining molecules like chlorophyll.

TABLE 5.2
Common NMR Techniques Used for the Identification of Natural Products

Technique	Mechanism	Note	Reference
1D NMR			
^{1}H	Shielding of ^{1}H nuclei	Functional groups are determined using chemical shifts, coupling allows the identification of protons on adjacent carbons and integration shows the number of identical protons	Pesek et al. (2020)
^{13}C	Shielding of ^{13}C nuclei	Functional groups are determined using chemical shifts	
2D NMR			
^{1}H-^{1}H Correlation spectroscopy (COSY)	Multiple radiofrequency pulses	Gives a much clearer indication of which protons are adjacent compared to coupling. Useful for identifying coupling in larger molecules	Fischer et al. (1991)
^{13}C Distortionless enhancement by polarisation transfer (DEPT)	Radiofrequency pulses from different angles	CH and CH$_3$ appear at 135° pulse angle. CH$_2$ appears at 90° pulse angle and all C bonded to H appears at 45° pulse angles	
^{1}H-^{1}H Nuclear overhauser effect spectroscopy (NOESY)	Three radiofrequency pulses 90° apart	Shows proximity of protons in space Useful for determining the shape of larger molecules	

TABLE 5.3
Commonly Used Column Packing Material

Column Packing Material	Primary Selectivity	Secondary Selectivity	Reference
Normal phase adsorption: i. Silica gel 60 ii. Alumina	Hydrophilicity	Molecular shape	Tan et al. (2015), Zhang et al. (2018), and Lim et al. (2021)
Reversed-phase partition: i. C18 ii. ODS iii. C8	Hydrophobicity		Tan et al. (2015), Gamal et al. (2019), and Lim et al. (2021)
Reversed-phase adsorption: i. MCI Gel CHP20P ii. Diaion HP20 iii. Sepabeads SP825L iv. Sepabeads SP850	Hydrophobicity	Molecular shape Pi-bond interaction	Le et al. (2017) and Roué et al. (2018)

(Continued)

TABLE 5.3 (*Continued*)
Commonly Used Column Packing Material

Column Packing Material	Primary Selectivity	Secondary Selectivity	Reference
Size exclusion:	Molecular size		Kawahara et al. (2019)
i. Sephadex G25			and Wang et al. (2021)
ii. Toyopearl HW40F			
iii. Bio-Gel P			
iv. Bio-Gel A			
Lipophilic size exclusion:	Molecular size	Hydrophobicity	Tan et al. (2015), Kholif
i. Sephadex LH20			et al. (2021), and Lim
ii. Bio-Beads SX			et al. (2021)
Ion exchange:	Ionic charge		Chand et al. (2020)
i. DEAE Sephadex A-50			
ii. CM Sephadex C50			

Similar fractions are pooled and subjected to bioactivity testing after each round of chromatography, and this process is repeated until the compound is pure (Tan et al., 2015; Lim et al., 2021). Eluted fractions are often monitored using thin-layer chromatography with a suitable visualising reagent such as a sulphuric acid spray or iodine vapour. Injecting fractions into HPLC is another common mistake as the process is prohibitively time-consuming with a large number of fractions and is often limited to UV detection.

5.4.3 Verifying Activity

One of the major hurdles, when associated with verifying activity, is that fractions may lose their activity as they are purified. There can be multiple reasons for this such as synergy between different compounds, the active compound(s) not being a major component of the fraction or the active compound becoming insoluble in aqueous solutions after purification. The latter frequently occurs with hydrophobic extracts, but aqueous solubility is an important factor in many bioassays. Compounds in a fraction may form eutectic mixtures or eutectic colloids, which would enhance their ability to disperse in water, and purification would prevent such mixtures (Bazzo et al., 2020). Loss of solubility can easily be compensated using surfactants.

Identifying IMPs and PAINs is also an important consideration when it comes to verifying activity. Computational chemistry using molecular docking and molecular dynamics simulation can help identify IMPs and PAINs. Molecular docking is a key tool in computer-assisted drug design that predicts how a compound binds to a target protein. After a compound's time-consuming isolation and structural elucidation, the compound can be fed into a docking software such as AutoDock (Österberg et al., 2002). This is, of course, limited to protein structures that are readily available in the database. Elucidating protein structures is another time-consuming process involving isolation, crystallisation and X-ray diffraction.

Molecular dynamics simulation of physical movements of atoms and molecules is often done to examine the interaction between smaller molecules such as isolated compounds with solvents and enzyme substrates. Molecular dynamics simulation has been used to study how PAINs can interact non-specifically with lipid bilayers (Magalhães et al., 2022).

Molecular dynamics can also be used to study uncompetitive inhibition from "crowding" where the PAIN molecule prevents the substrate from physically interacting with the enzyme (Matić et al., 2020). Many PAINs are also aggregators which cause the molecules to precipitate out of the solution (Bisson et al., 2016). Uncompetitive inhibition should be distinguished from non-competitive inhibition, where the inhibitor molecules bind specifically to an enzyme's allosteric site.

PAINs can be readily identified using computational chemistry and this would significantly reduce the number of IMPs. However, IMPs that are not PAINs may be harder to identify, and even then, they still attract the attention of scientists that are not familiar with the concept of IMPs and PAINs. IMPs can also be identified using cell, animal and clinical models. This is the case with hydroxycitric acid which had many controversial clinical trials of dubious efficacy (Cruz et al., 2021). As things are, many IMPs and PAINs remain unidentified as natural product chemistry and computational chemistry require very different skill sets.

5.5 BIOACTIVITY TESTING

5.5.1 Antioxidant Activity

Excessive oxidative stress is tied to ageing, cardiovascular disease, cancer, diabetes and neurodegenerative diseases such as Alzheimer's disease, Parkinson's disease and multiple sclerosis (Tan and Lim, 2015). This has generated a lot of interest in the development of antioxidant assays (Ng et al., 2021).

Common antioxidant assays are shown in Table 5.4. They range from antioxidant content which is selective of specific antioxidant compounds in plants to various primary antioxidant and secondary antioxidant activities. Primary antioxidants involve direct scavenging of radicals and examples include lipid peroxidation inhibition, hydrophobic radical scavenging, oxygen radical scavenging and metal ion reduction. Secondary antioxidants are those which prevent the generation of free radicals and the most common example is metal ion chelation which prevents the generation of hydroxide radicals by the Fenton reaction.

Many of these assays in Table 5.4 are conducted using non-enzymatic and cell-free reagents but some can be adapted for use in enzymatic and cell-based models. Ferric reducing ability of plasma (FRAP) and thiobarbituric acid reactive substances (TBARS) are used to assess antioxidants in plasma and they can also be adapted to assess the inhibition of pro-oxidant enzymes such as lipoxygenase and hyaluronidase (Marchelak et al., 2017). The oxygen radical absorbance capacity (ORAC) assay can be adapted for studying intracellular oxygen radicals by using the dichlorodihydrofluorescein diacetate dye which is more cell permeable than fluorescein (Dikalov and Harrison, 2014).

TABLE 5.4

Common Antioxidant Assays

Classification	Assay	Common Standard	Note	Reference
Antioxidant content	Folin-Ciocalteu	Gallic acid	Sodium tungstate and phosphomolybdic acid react with any reducing substances, which in plants are mostly phenols. Sugars and ascorbic acid are interfering compounds, but these are only a problem in aqueous extracts	Tan and Lim (2015)
	Potassium acetate/aluminium chloride	Quercetin	Aluminium chloride will react with flavonoids to form a yellow-coloured complex. More sensitive to flavonols like quercetin	Shraim et al. (2021)
	Sodium nitrite/aluminium chloride	Catechin	A variation of the aluminium chloride assay with the addition of sodium nitrite as an oxidising agent. Complexing with aluminium chloride after oxidation gives a red colour. More sensitive to flavanols like catechin	Shraim et al. (2021)
	Sodium molybdate	Chlorogenic acid	Caffeoylquinic acids complex with molybdate, an ion, to form a yellowish complex	Tan et al. (2015)
Lipid peroxidation inhibition	β-Carotene bleaching	Quercetin	Determines the ability of antioxidants such as quercetin in breaking the lipid peroxidation chain reaction triggered by heating linoleic acid in an aqueous emulsion. β-carotene is added as an indicator and need not be present in the plant sample. Some antioxidants such as ascorbic acid would not break the chain reaction but would instead perpetuate it	Chan et al. (2009)
	TBARS	Malondialdehyde	Used to measure lipid peroxidation in serum and cell lysate. Endoperoxides from serum and cell lysates oxidise fatty acids which would react with thiobarbituric acid to form the pinkish malondialdehyde. Antioxidants will reduce malondialdehyde formation	Tan and Lim (2015)
Hydrophobic radical scavenging	DPPH scavenging	Ascorbic acid, Trolox	DPPH is a hydrophobic, purple-coloured free radical which is often used to measure radical scavenging in plants. The scavenging capacity does not correlate linearly with concentration, and as such, the activity is often expressed as IC_{50}, instead of percent scavenging. Ascorbic acid and Trolox are used as standards to compare with hydrophilic and lipophilic antioxidants, respectively	Tan and Lim (2015)

(Continued)

TABLE 5.4 (Continued)
Common Antioxidant Assays

Classification	Assay	Common Standard	Note	Reference
	ABTS scavenging	Ascorbic acid Trolox	ABTS is oxidised into a hydrophobic ABTS+ radical which is green in colour which would be scavenged by antioxidants from plants. Like DPPH, the scavenging capacity does not correlate linearly with the concentration, and as such, the activity is often expressed as IC_{50} instead of percent scavenging. Ascorbic acid and Trolox are used as standards to compare with hydrophilic and lipophilic antioxidants, respectively.	Tan and Lim (2015)
Oxygen radical scavenging	ORAC	Trolox	Peroxyl radicals are generated using 2'-azobis (2-amidino-propane) dihydrochloride. These would oxidise the fluorescein into non-fluorescent products. Antioxidants such as Trolox would prevent the oxidation of fluorescein	Tan and Lim (2015)
	NBT reduction	Quercetin N-acetylcysteine	Superoxide radicals from the non-enzymatic NADH/PMS or cells such as leucocytes reduce NBT into a blue-black formazan dye. Antioxidants such as quercetin and N-acetylcysteine would reduce formazan formation	Tan and Lim (2015)
Metal ion reduction	Potassium ferricyanide	Gallic acid	Potassium ferricyanide would be reduced by antioxidants into potassium ferrocyanide that would in turn react with Fe^{3+} ions to form a Prussian blue complex. This assay is used to measure ferric reduction in a non-enzymatic system	Tan et al. (2015)
	FRAP	Trolox	Fe (III)/tripyridyltriazine complex is used to measure the antioxidant capacity of plasma which would reduce it to a blue Fe (II) form. It is used to study antioxidant supplementation with plant compounds. It is often mixed up with the similar potassium ferricyanide assay which sometimes shares the same FRAP abbreviation	Tan and Lim (2015)
	CUPRAC	Trolox	Similar to the potassium ferricyanide assay. Uses Cu (II) Bis(neocuproine) light blue complex which is reduced to Cu (I) to give a yellow colour	Tan and Lim (2015)
Metal ion chelation	Ferrozine	EDTA	Ferrozine is a ligand which forms a purple complex with Fe (II). Chelating agents that can out-compete ferrozine are considered secondary metabolites	Tan et al. (2015)

Abbreviations: TBARS, Thiobarbituric acid reactive substances; DPPH, 2,2-diphenyl-1-picrylhydrazyl; ABTS, 2,2'-azino-bis(3-ethylbenzothiazoline-6-sulphonic acid); ORAC, Oxygen radical absorbance capacity; FRAP, Ferric ion reducing antioxidant power/Ferric reducing ability of plasma; CUPRAC, Cupric ion reducing antioxidant capacity.

5.5.2 Cytotoxicity, Apoptosis and Inflammation

Although closely related to oxidative stress, nonetheless, bioactivity testing for inflammation, apoptosis and cytotoxicity (often within the context of anticancer activity) differs considerably from in vitro antioxidant activity assays due to the obvious necessity for cell culture as opposed to the often cell-free antioxidant assays (Fink and Cookson, 2005). The MTT (3-(4,5-Dimethylthiazol-2-yl)-2,5-diphenyltetrazolium bromide) and MTS (3-(4,5-di-methylthiazol-2-yl)-5-(3-carboxymethoxyphenyl)-2-(4-sulphophenyl)-2H-tetrazolium) assays are widely used for the screening of cytotoxicity and anticancer activity of natural compounds, where both measure mitochondrial metabolism as an indirect measure of viable cell numbers.

Cell proliferation on the other hand can be measured with the BrdU cell proliferation enzyme-linked immunosorbent assay (Yew et al., 2019). There are also many assays that either measure ATP production (e.g. luminescence luciferase-based assays) or DNA production (e.g. fluorescent DNA-binding dyes). In some cases, these assays may be better-suited than MTT- or MTS-based assays for measuring cell viability as several compounds (notably phenolics) may directly reduce these tetrazolium salts to formazan and/or increase mitochondrial succinate dehydrogenase activity (Wang et al., 2010).

Cells that are subjected to cytotoxic assays or cell proliferation assays are often examined under the microscope for signs of apoptosis and necrosis (Fink and Cookson, 2005). Apoptosis or programmed cell death does not elicit an inflammatory response. In contrast, necrosis or cell death occurs from environmental perturbations that would release inflammatory cellular contents. The former is preferred for cytotoxic therapies. However, with cell proliferation assays, cell growth may be inhibited without cell death, indicated by the absence of apoptotic or necrotic cells.

In the absence of cell culture facilities, brine shrimp lethality assays are sometimes used as a preliminary test for cytotoxicity (Ng et al., 2021). Brine shrimp has the advantage of requiring very simple reagents and apparatus, i.e. shrimp eggs, salt water and air stone. However, its main drawback is that toxicity in shrimps would not reflect those in mammalian cells. Furthermore, the assay does not allow the study of the mode of cell death and inflammation and is limited only to hydrophilic substances. Organic solvents can kill brine shrimp, and care must be taken when using organic solvents to help with sample solubility.

5.5.3 Antimicrobial Activity

A major role of secondary metabolites is the protection against microbial pathogens, be they bacterial, viral or fungal. As would be expected, many natural compounds exhibit some degree of inhibitory or lethal activity against microbes. Screening of antibacterial and antifungal activity can be readily accomplished through a wide range of susceptibility tests which are relatively rapid and of low cost.

Common assays include disk diffusion (Bibi et al., 2021), agar diffusion and broth dilution (Ma et al., 2018), potentially in line with Clinical and Laboratory Standards Institute (CLSI) guidelines (Tan and Lim, 2015). This can be easily expanded upon by determining the effect of the extract or compound over time (e.g. time-kill kinetics)

(Bibi et al., 2021) or observing whether the effects are merely inhibitory (-static) or lethal (-cidal) to the test organisms (e.g. via flow cytometry with propidium iodide staining) (García-Varela et al., 2015).

5.6 INSTRUMENTATION

Analytical instruments have much in common with those of preparative separations (Schmidt-Traub et al., 2020). Analytical separation prioritises good resolution of peaks and is often conducted under high pressure. Preparative separations tend to be simple by applying gravity-based columns. Many of the chromatographic methods described earlier in Table 5.3, Section 5.4, that are used for preparative separation are also used for analytical separations with a few exceptions. Size-exclusion resins usually deform under pressure and thus are not used for analytical separation. Preparative and analytical separations differ in their goal which is summarised in Table 5.5. Analytical separations require purpose-built instruments to ensure reproducible retention times and signal response for quantitative analysis.

Given that preparative separations have been well discussed in Section 5.4, this section would focus on analytical separations, which are mostly chromatographic methods with different detectors for identification and quantification. Chromatographic separations are classified according to their mobile phase, i.e. liquid and gas chromatography. The former is also used for preparative separations, while the latter is almost exclusively used in analytical separations.

Theoretical plates are typically used as a measure of the resolution and efficiency of chromatography columns (Agilent, 2012; Sigma-Aldrich, 2013). Liquid

---

TABLE 5.5

Analytical versus Preparative and Semipreparative

Item	Preparative	Semipreparative	Analytical
Scale	300–1,000 g of packing	100–1 g of packing	1 g or less of packing
Sample load	10 g or more of sample	10 g–10 mg of sample	Usually in the ng range
Resolution	Low	Medium	High
Reproducibility	Low	Medium	High
Detector sensitivity	Low	Medium	High
Speed	Low	Medium	High
Priority	High sample load to isolate sufficient quantities for bioassays. Speed is not a priority, conducted over days	High sample purity for structural elucidation and verifying activity. Speed is not a priority, conducted over hours	High sensitivity, low sample load but reproducible retention times and signal response for quantitative analysis. Speed is also important when analysing many samples. Conducted in minutes

Source: Schmidt-Traub et al. (2020).

---

chromatography (LC) columns typically have about 5,000 plates while gas chromatography (GC) can have more than 100,000 plates. In general, GC is superior for analytical separations with peaks appearing as thin lines, with much higher sensitivity and a very reliable mass spectra database for compound identification. LC is superior in terms of versatility as non-volatile compounds cannot be analysed using GC.

5.6.1 Liquid Chromatography

LC uses a liquid as the mobile phase but the stationary phase can be either liquid or solid in the case of partition and adsorption stationary phases, respectively. Liquid chromatographs are probably the most widely used analytical instruments for quantitative analysis as they can be used for a wide range of samples (Cavaliere et al., 2018).

The solvent pump is probably the most important distinguishing factor between analytical and preparative separations (Swartz, 2005). The pump is required to move the mobile phase at a higher speed than possible with gravity columns. This reduces band broadening from passive diffusion which is a serious problem at low flow rates and improves resolution. High pressure also allows the use of smaller particle size packing material which would drastically increase resolution. This is such a substantial increase in resolution that analytical chromatographs are classified according to pressure:

i. HPLC which typically runs at 100 bar with 3–5-micron packing
ii. Ultra-performance liquid chromatography (UPLC) which typically runs at 1,000 bar with sub-2-micron packing

UPLC is a much more costly instrument but it offers separation, which is on par with GC and in a much shorter time with high pressure and flow rates used (Swartz, 2005). Besides this important upgrade, HPLC and UPLC also serve similar purposes. The stationary phase is the next most important component. Many of these are described in Table 5.3, but examples of stationary phases used in food samples are listed in Table 5.6.

TABLE 5.6
Examples of Liquid Chromatography Used in Foods

Sample	Bioactive Compound	Stationary Phase	Mobile Phase	References
Nigella sativa	Small proteins about 94 ± 10 kDa in mass	Anion exchange DEAE Sephadex A50	0.05 M phosphate buffer (pH 6.4) containing 0.01 M NaCl	Haq et al. (1999)
Soybean	Globulin and glycinin	Anion exchange POROS HQ/10	Buffer solution gradient from pH 7 to 10	Heras et al. (2007)
Spinach leaves	Chlorophylls and carotenoids	Reversed-phase C18 column	Aqueous ethanol gradient	Reese (1997)
Costus speciosus	Phenolics	Reversed-phase C18 column	Aqueous acetonitrile gradient	Karthikeyan et al. (2012)

TABLE 5.7
Detectors Used Together with Modern HPLC Systems

Detector	Selectivity	Sensitivity	Structural Information	Reference
Photodiode array	UV chromophores, especially aromatic rings	ppb	Chromophores distinguished based on UV spectra	Jalaludin and Kim (2021)
Refractive index	Organic compounds	ppm	No structural information	Jalaludin and Kim (2021)
Electrospray ionisation (ESI-MS)	Hydrophilic organic compounds	ppb	Molar mass	Souverain et al. (2004)
Atmospheric pressure chemical ionisation (APCI-MS)	Hydrophobic organic compounds	ppb	Molar mass	Souverain et al. (2004)
^{1}H NMR	Organic compounds	ppm	^{1}H chemical shifts	Lhoste et al. (2022)

5.6.2 Liquid Chromatography Detectors

Detectors are sensitive to different analytes and can provide different information about the molecule. Detectors can range from general-purpose photodiode array detectors to specific detectors such as mass spectrometers as summarised in Table 5.7.

Photodiode arrays are by far the most common detector used with HPLC systems due to their simple instrumentation, high sensitivity and wide selectivity (Jalaludin and Kim, 2021). However, refractive index detectors, despite suffering from low sensitivity, are often necessary for compounds without aromatic rings or double bonds such as sugars.

Electrospray ionisation-MS (ESI-MS) and atmospheric pressure chemical ionisation-MS (APCI-MS) are less common because of their higher maintenance cost. Both MS detectors are similar, with the latter being able to ionise more hydrophobic compounds using an electrical arc (Liigand et al., 2021). Given that ESI and APCI are soft ionisation methods, they are often used on tandem MS systems that promote fragmentation and can be matched with mass databases. However, identification using mass databases for LC is still quite unreliable, and confirmation with NMR is still necessary.

NMR-based detectors are by far the least common despite their impressive ability to rapidly produce NMR spectra for compound identification (Lhoste et al., 2022). The rarity for most of the parts is due to the high cost. The instruments are still limited to ^{1}H because of the scarcity of ^{13}C isotopes in nature. Furthermore, online integration of LC and NMR results in low sensitivity and limited ability to produce 2D NMR spectra, but recent advances are pushing further these limits.

5.6.3 Gas Chromatography

GC uses gas as the mobile phase and a liquid-bonded phase within a capillary column which often reaches 30 m in length. Like LC, there are many stationary phases available for GC as outlined in Table 5.8.

TABLE 5.8
Common GC Stationary Phases

Stationary Phase	Polarity	Max. Temperature (°C)	Reference
Squalene	Non-polar	150	Mota et al. (2021)
Methylpolysiloxane		300	
Phenylmethylpolysiloxane		300	
Cyanopropylphenyl	Intermediate	300	Engewald et al. (2014)
Methylpolysiloxane		225	
Phenylmethylpolysiloxane		370	
Phenylsiloxane carborane			
Carbowax (Polyethylene glycol)	Polar	225	McNair et al. (2019)
Cyanopropyl methylpolysiloxane		275	

Despite having many stationary phases to choose from, the choice of stationary phase is not the most important consideration in GC (McNair et al., 2019). The boiling points of analyte molecules mostly determine the separation in GC. Stationary phases are matched with analytes according to the polarity, and the temperature used should not exceed those specified for the stationary phase. Similarly, the choice of mobile phase also does not affect selectivity. Elution is dependent on the operating temperature of GC, relative to the boiling points of the analytes. Hydrogen gas is the preferred mobile phase because of its high diffusivity.

5.6.4 GAS CHROMATOGRAPHY DETECTORS

GC instruments have much fewer detectors as compared to LC and these detectors are outlined in Table 5.9. The choice of detectors in GC is quite straightforward and Flame ionisation detectors are the go-to for quantitative analysis. Electron impact-MS is a hard ionisation method that does not require a tandem MS. Mass databases for GC are much more reliable than those for LC-MS. Olfactory detectors have the most niche application as they use the operator's nose as a qualitative detector to describe flavour molecules that are elected from the column. Olfactory detectors often comprise an olfactory port for the operator to smell compounds as they are eluted.

TABLE 5.9
Detectors Used Together with Modern HPLC Systems

Detector	Selectivity	Sensitivity	Structural Information	Reference
Flame ionisation (FID)	Carbon compounds	ppb	None	McNair et al. (2019)
Electron impact (EI-MS)	Carbon compounds	ppb	Mass fragments matched to reliable mass databases	
Olfactory	Flavour molecules	Qualitative	None	

The technology associated with GC is mature compared to LC, and what it lacks in diversity, it more than makes up for in terms of sensitivity, reliability and simplicity (McNair et al., 2019). The main limitation of GC is that its use is limited to volatile compounds. Profiling essential oils from plants using GC-MS is considered routine.

5.7 CONCLUSION AND FUTURE DEVELOPMENT

Isolating, identifying and analysing bioactive compounds in herbs, spices and medicinal plants are multidisciplinary endeavours, and it is very much an art as it is a science. It is an art as practitioners need a superficial understanding of a wide range of concepts related to chemistry, physics and biology. However, the technology available is mature and mastery of different technological facets is very much a science. Engineers and chemists are consistently pushing the limits of sensitivity, precision and speed of analysis.

Many aspects of natural product analysis are becoming routine and some groups of natural products run the risk of them being overstudied such as hydroxycitric acid from *Garcinia cambogia* and curcumin from *Curcuma longa*. There are many repeated studies that are not meaningfully different and require heavy investment into natural product drug leads, which may turn out to be either an IMP or a PAIN. Hence, greater effort must be placed into verifying activity through computational chemistry, cell-based models and clinical studies. This, however, requires familiarity with an even wider skill set.

REFERENCES

Agilent. (2012). *LC AND LC/MS*. Agilent Technologies.

Atanasov, A. G., Zotchev, S. B., Dirsch, V. M., & Supuran, C. T. (2021). Natural products in drug discovery: Advances and opportunities. *Nature Reviews Drug Discovery*, *20*(3), 200–216.

Bajorath, J. (2021). Evolution of assay interference concepts in drug discovery. *Expert Opinion on Drug Discovery*, *16*(7), 719–721.

Batsis, J. A., Apolzan, J. W., Bagley, P. J., Blunt, H. B., Divan, V., Gill, S., … & Kidambi, S. (2021). A systematic review of dietary supplements and alternative therapies for weight loss. *Obesity*, *29*(7), 1102–1113.

Bazzo, G. C., Pezzini, B. R., & Stulzer, H. K. (2020). Eutectic mixtures as an approach to enhance solubility, dissolution rate and oral bioavailability of poorly water-soluble drugs. *International Journal of Pharmaceutics*, *588*, 119741.

Bibi, M., Murphy, S., Benhamou, R. I., Rosenberg, A., Ulman, A., Bicanic, T., Fridman, M., & Berman, J. (2021). Combining colistin and fluconazole synergistically increases fungal membrane permeability and antifungal cidality. *ACS Infectious Diseases*, *7*(2), 377–389.

Biradi, M., & Hullatti, K. (2017). Bioactivity-guided isolation of cytotoxic terpenoids and steroids from *Premna serratifolia*. *Pharmaceutical Biology*, *55*(1), 1375–1379.

Bisson, J., McAlpine, J. B., Friesen, J. B., Chen, S. N., Graham, J., & Pauli, G. F. (2016). Can invalid bioactives undermine natural product-based drug discovery?. *Journal of Medicinal Chemistry*, *59*(5), 1671–1690.

Casciaro, B., Calcaterra, A., Cappiello, F., Mori, M., Loffredo, M. R., Ghirga, F., Mangoni, M. L., Botta, B., & Quaglio, D. (2019). Nigritanine as a new potential antimicrobial alkaloid for the treatment of *Staphylococcus aureus*-induced infections. *Toxins*, *11*, 511.

Cavaliere, C., Capriotti, A.L., La Barbera, G., Montone, C.M., Piovesana, S., Laganà, A. (2018). Liquid chromatographic strategies for separation of bioactive compounds in food matrices. *Molecules*, *23*(12), 3091. https://doi.org/10.3390/molecules23123091

Chan, E. W. C., Lim, Y. Y., & Chew, Y. L. (2007). Antioxidant activity of Camellia sinensis leaves and tea from a lowland plantation in Malaysia. *Food Chemistry*, *102*(4), 1214–1222.

Chan, E. W. C., Lim, Y. Y., Wong, S. K., Lim, K. K., Tan, S. P., Lianto, F. S., & Yong, M. Y. (2009). Effects of different drying methods on the antioxidant properties of leaves and tea of ginger species. *Food Chemistry*, *113*(1), 166–172.

Chan, E. W. C., Chan, H. J., Lim, J. E., Yik, S. H., Tan, S. F., Yap, P. Y., Goh, P.C., & Yee, S. Y. (2015). Effects of different cooking methods on the bioactivities of some spices. *Emirates Journal of Food and Agriculture*, *27*(8), 610–616.

Chan, E. W. C., Ong, A. C. L., Lim, K. L., Chong, W. Y., Chia, P. X., & Foo J. P. Y. (2019). Effects of superheated steam drying on the antioxidant and anti-tyrosinase properties of selected *Labiatae* herbs. *Carpathian Journal of Food Science & Technology*, *11*(1), 166–177.

Chand, S., Mahajan, R. V., Prasad, J. P., Sahoo, D. K., Mihooliya, K. N., Dhar, M. S., & Sharma, G. (2020). A comprehensive review on microbial l-asparaginase: Bioprocessing, characterisation, and industrial applications. *Biotechnology and Applied Biochemistry*, *67*(4), 619–647.

Cheng, H. S., Goh, B. H., Phang, S. C. W., Amanullah, M. M., Ton, S. H., Palanisamy, U. D., Abdul Kadir, K., & Tan, J. B. L. (2020). Pleiotropic ameliorative effects of ellagitannin geraniin against metabolic syndrome induced by high-fat diet in rats. *Nutrition*, *80*, 110973.

Cruz, A. C., Pinto, A. H., Costa, C. D., Oliveira, L. P., Oliveira-Neto, J. R., & Cunha, L. C. (2021). Food-effect on (−)–hydroxycitric acid absorption after oral administration of *Garcinia cambogia* extract formulation: A phase I, randomized, cross-over study. *Journal of Pharmaceutical Sciences*, *110*(2), 693–697.

Das, K., & Gezici, S. (2018). Secondary plant metabolites, their separation and identification, and role in human disease prevention. *Annals of Phytomedicine*, *7*, 13–24.

Dikalov, S. I., & Harrison, D. G. (2014). Methods for detection of mitochondrial and cellular reactive oxygen species. *Antioxidants & Redox Signaling*, *20*(2), 372–382.

Engewald, W., Dettmer-Wilde, K., & Rotzsche, H. (2014). Columns and stationary phases. In Katja Dettmer-Wilde and Werner Engewald (Eds.), *Practical Gas Chromatography* (pp. 59–116). Springer, Berlin, Heidelberg.

Ferreira, I. C., Martins, N., & Barros, L. (2017). Phenolic compounds and its bioavailability: In vitro bioactive compounds or health promoters? In Fidel Toldrá (Ed.), *Advances in Food and Nutrition Research* (Vol. 82, pp. 1–44). Academic Press.

Fink, S. L., & Cookson, B. T. (2005). Apoptosis, pyroptosis, and necrosis: Mechanistic description of dead and dying eukaryotic cells. *Infection and Immunity*, *73*(4), 1907–1916.

Fischer, N. H., Vargas, D., & Menelaou, M. (1991). Modern NMR methods in phytochemical studies. In Nikolaus H. Fischer, Murray B. Isman, and Helen A. Stafford (Eds.), *Modern Phytochemical Methods* (pp. 271–317). Springer, Boston, MA.

Gamal, M., Ali, H. M., Abdelfatah, R. M., & Magdy, M. A. (2019). A green approach for simultaneous analysis of two natural hepatoprotective drugs in pure forms, capsules and human plasma using HPLC-UV method. *Microchemical Journal*, *151*, 104258.

García-Varela, R., García-García, R. M., Barba-Dávila, B. A., Fajardo-Ramírez, O. R., Serna-Saldívar, S. O., & Cardineau, G. A. (2015). Antimicrobial activity of *Rhoeo discolor* phenolic rich extracts determined by flow cytometry. *Molecules*, *20*(10), 18685–18703.

Hajduk, P. J., & Greer, J. (2007). A decade of fragment-based drug design: Strategic advances and lessons learned. *Nature Reviews Drug discovery*, *6*(3), 211–219.

Haq, A., Lobo, P. I., Al-Tufail, M., Rama, N. R., & Al-Sedairy, S. T. (1999). Immunomodulatory effect of Nigella sativa proteins fractionated by ion exchange chromatography. *International Journal of Immunopharmacology, 21*(4), 283–295.

Heras, J., Marina, M., & Garcia, M. (2007). Development of a perfusion ion-exchange chromatography method for the separation of soybean proteins and its application to cultivar characterisation. *Journal of Chromatography A, 1153*(1–2), 97–103.

Inada, M., Shindo, M., Kobayashi, K., Sato, A., Yamamoto, Y., Akasaki, Y., Ichimura, K., & Tanuma, S. I. (2019). Anticancer effects of a non-narcotic opium alkaloid medicine, papaverine, in human glioblastoma cells. *PLoS ONE, 14*(5), 1–9.

Jalaludin, I., & Kim, J. (2021). Comparison of ultraviolet and refractive index detections in the HPLC analysis of sugars. *Food Chemistry, 365*, 130514.

Karthikeyan, J., Reka, V., & Giftson, R. V. (2012). Characterisation of bioactive compounds in Costus speciosus (Koen). by reverse phase HPLC. *International Journal of Pharmaceutical Sciences and Research, 3*(5), 1461.

Kawahara, S. I., Ishihara, C., Matsumoto, K., Senga, S., Kawaguchi, K., Yamamoto, A., ... & Fujii, H. (2019). Identification and characterisation of oligomeric proanthocyanidins with significant anti-cancer activity in adzuki beans (*Vigna angularis*). *Heliyon, 5*(10), e02610.

Kholif, O. T., Sebaei, A. S., Eissa, F. I., & Elhamalawy, O. H. (2021). Size-exclusion chromatography selective cleanup of aflatoxins in oilseeds followed by HPLC determination to assess the potential health risk. *Toxicon, 200*, 110–117.

Le, C. F., Kailaivasan, T. H., Chow, S. C., Abdullah, Z., Ling, S. K., & Fang, C. M. (2017). Phytosterols isolated from *Clinacanthus nutans* induce immunosuppressive activity in murine cells. *International Immunopharmacology, 44*, 203–210.

Lešková, E., Kubíková, J., Kováčiková, E., Košická, M., Porubská, J., & Holčíková, K. (2006). Vitamin losses: Retention during heat treatment and continual changes expressed by mathematical models. *Journal of Food Composition and Analysis, 19*(4), 252–276.

Lhoste, C., Lorandel, B., Praud, C., Marchand, A., Mishra, R., Dey, A., ... & Giraudeau, P. (2022). Ultrafast 2D NMR for the analysis of complex mixtures. *Progress in Nuclear Magnetic Resonance Spectroscopy, 130–131*, 1–46.

Liigand, P., Liigand, J., Kaupmees, K., & Kruve, A. (2021). 30 Years of research on ESI/MS response: Trends, contradictions and applications. *Analytica Chimica Acta, 1152*, 238117.

Lim, C. S., Chan, E. W., Wong, C. W., Tan, J. B., Anggraeni, V. S., Loong, Z. J., & Teo, Y. K. (2021). Anti-quorum sensing and antibiotic enhancement of allylpyrocatechol and methyl gallate. *Jurnal Teknologi, 83*(3), 101–106.

Ma, M., Wen, X., Xie, Y., Guo, Z., Zhao, R., Yu, P., ... & Zeng, Z. (2018). Antifungal activity and mechanism of monocaprin against food spoilage fungi. *Food Control, 84*, 561–568.

Magalhães, P. R., Reis, P. B., Vila-Viçosa, D., Machuqueiro, M., & Victor, B. L. (2022). Optimization of an in silico protocol using probe permeabilities to identify membrane pan-assay interference compounds. *Journal of Chemical Information and Modeling, 62*(12), 3034–3042.

Marchelak, A., Owczarek, A., Matczak, M., Pawlak, A., Kolodziejczyk-Czepas, J., Nowak, P., & Olszewska, M. A. (2017). Bioactivity potential of *Prunus spinosa* L. flower extracts: Phytochemical profiling, cellular safety, pro-inflammatory enzymes inhibition and protective effects against oxidative stress in vitro. *Frontiers in Pharmacology, 8*, 680.

Martinsen, B. K., Aaby, K., & Skrede, G. (2020). Effect of temperature on stability of anthocyanins, ascorbic acid and color in strawberry and raspberry jams. *Food Chemistry, 316*, 126297.

Matić, M., Saurabh, S., Hamacek, J., & Piazza, F. (2020). Crowding-induced uncompetitive inhibition of lactate dehydrogenase: Role of entropic pushing. *The Journal of Physical Chemistry B, 124*(5), 727–734.

McNair, H. M., Miller, J. M., & Snow, N. H. (2019). *Basic Gas Chromatography*. John Wiley & Sons, Hoboken, NJ.

Montenegro, Z. J. S., Álvarez-Rivera, G., Sánchez-Martínez, J. D., Gallego, R., Valdés, A., Bueno, M., Cifuentes A., & Ibáñez, E. (2021). Neuroprotective effect of terpenoids recovered from olive oil by-products. *Foods, 10*(7), 1507.

Mota, M. F. S., Waktola, H. D., Nolvachai, Y., & Marriott, P. J. (2021). Gas chromatography–mass spectrometry for characterisation, assessment of quality and authentication of seed and vegetable oils. *TrAC Trends in Analytical Chemistry, 138*, 116238.

Ng, Z. X., & Kuppusamy, U. R. (2019). Effects of different heat treatments on the antioxidant activity and ascorbic acid content of bitter melon, *Momordica charantia. Brazilian Journal of Food Technology, 22*, e2018283.

Ng, Z. X., Samsuri, S. N., & Yong, P. H. (2020a). The antioxidant index and chemometric analysis of tannin, flavonoid, and total phenolic extracted from medicinal plant foods with the solvents of different polarities. *Journal of Food Processing and Preservation, 44*(9), e14680.

Ng, Z. X., Yong, P. H., & Lim, S. Y. (2020b). Customized drying treatments increased the extraction of phytochemicals and antioxidant activity from economically viable medicinal plants. *Industrial Crops and Products, 155*, 112815.

Ng, Z. X., Koick, Y. T. T., & Yong, P. H. (2021). Comparative analyses on radical scavenging and cytotoxic activity of phenolic and flavonoid content from selected medicinal plants. *Natural Product Research, 35*(23), 5271–5276.

Österberg, F., Morris, G.M., Sanner, M.F., Olson, A.J. and Goodsell, D.S. (2002), Automated docking to multiple target structures: Incorporation of protein mobility and structural water heterogeneity in AutoDock. *Proteins 46*, 34–40. https://doi.org/10.1002/prot.10028

Pesek, M., Juvan, A., Jakoš, J., Košmrlj, J., Marolt, M., & Gazvoda, M. (2020). Database independent automated structure elucidation of organic molecules based on IR, ^{1}H NMR, ^{13}C NMR, and MS Data. *Journal of Chemical Information and Modeling, 61*(2), 756–763.

Ramawat, K. G., Dass, S., & Mathur, M. (2009). The chemical diversity of bioactive molecules and therapeutic potential of medicinal plants. In *Herbal Drugs: Ethnomedicine to Modern Medicine* (pp. 7–32). K. G. Ramawat (Ed.). Springer Berlin, Heidelberg.

Reese, R. N. (1997). Separation of chloroplast pigments using reverse phase chromatography. *American Biology Teacher, 59*(2), 114–117.

Roué, M., Darius, H. T., & Chinain, M. (2018). Solid phase adsorption toxin tracking (SPATT) technology for the monitoring of aquatic toxins: A review. *Toxins, 10*(4), 167.

Schmidt-Traub, H., Schulte, M., & Seidel-Morgenstern, A. (Eds.). (2020). *Preparative Chromatography*. John Wiley & Sons, Weinheim, Germany.

Shraim, A. M., Ahmed, T. A., Rahman, M. M., & Hijji, Y. M. (2021). Determination of total flavonoid content by aluminum chloride assay: A critical evaluation. *LWT, 150*, 111932.

Sigma-Aldrich. (2013). *GC Column Selection Guide*. Sigma-Aldrich Co, Buckinghamshire, UK.

Siow, C. S., Chan, E. W. C., Wong, C. W., & Ng, C. W. (2022). Antioxidant and sensory evaluation of cocoa (*Theobroma cacao* L.) tea formulated with cocoa bean hull of different origins. *Future Foods, 5*, 100108.

Snyder, L. R. (1974). Classification of the solvent properties of common liquids. *Journal of Chromatography A, 92*(2), 223–230.

Solinas, G., Borén, J., & Dulloo, A. G. (2015). De novo lipogenesis in metabolic homeostasis: More friend than foe? *Molecular Metabolism, 4*(5), 367–377.

Soong, Y.-Y., & Barlow, P. J. (2006). Quantification of gallic acid and ellagic acid from longan (*Dimocarpus longan* Lour.) seed and mango (*Mangifera indica* L.) kernel and their effects on antioxidant activity. *Food Chemistry, 97*(3), 524–530.

Souverain, S., Rudaz, S., & Veuthey, J. L. (2004). Matrix effect in LC-ESI-MS and LC-APCI-MS with off-line and online extraction procedures. *Journal of Chromatography A*, *1058*(1–2), 61–66.

Swartz, M. E. (2005). UPLC™: An introduction and review. *Journal of Liquid Chromatography & Related Technologies*, *28*(7–8), 1253–1263.

Tan, J. B. L., & Lim, Y. Y. (2015). Critical analysis of current methods for assessing the in vitro antioxidant and antibacterial activity of plant extracts. *Food Chemistry*, *172*, 814–822.

Tan, J. J., Tan, J. B., Lee, S. M., & Lim, Y. Y. (2020). Antioxidant and antimicrobial potential of Costus woodsonii. *Journal of Herbs, Spices & Medicinal Plants*, *26*(2), 191–202.

Tan, Y. P., Chan, E. W. C., & Lim, C. S. Y. (2015). Potent quorum sensing inhibition by methyl gallate isolated from leaves of *Anacardium occidentale* L.(cashew). *Chiang Mai Journal of Science*, *42*(3), 650–656.

Tian, K. M., Li, J. J., & Xu, S. W. (2019). Rutaecarpine: A promising cardiovascular protective alkaloid from *Evodia rutaecarpa* (Wu Zhu Yu). *Pharmacological Research*, *141*(January), 541–550.

Troost, B., Mulder, L. M., Diosa-Toro, M., van de Pol, D., Rodenhuis-Zybert, I. A., & Smit, J. M. (2020). Tomatidine, a natural steroidal alkaloid shows antiviral activity towards chikungunya virus in vitro. *Scientific Reports*, *10*(1), 6364.

Vuolo, M. M., Lima, V. S., & Junior, M. R. M. (2019). Phenolic compounds: Structure, classification, and antioxidant power. In Maira Rubi Segura Campos (Ed.), *Bioactive Compounds* (pp. 33–50). Woodhead Publishing.

Wang, P., Henning, S. M., & Heber, D. (2010). Limitations of MTT and MTS-based assays for measurement of antiproliferative activity of green tea polyphenols. *PLoS ONE*, *5*(4), e10202.

Wang, G., Wang, H., Chen, Y., Pei, X., Sun, W., Liu, L., … & Wang, M. (2021). Optimisation and comparison of the production of galactooligosaccharides using free or immobilised *Aspergillus oryzae* β-galactosidase, followed by purification using silica gel. *Food Chemistry*, *362*, 130195.

Wong, S. K., Lim, Y. Y., & Chan, E. W. C. (2009). Antioxidant properties of *Hibiscus*: Species variation, altitudinal change, coastal influence and floral colour change. *Journal of Tropical Forest Science*, *21*, 307–315.

Xiu, F., Li, X., Zhang, W., He, F., Ying, X., & Stien, D. (2019). A new alkaloid from *Portulaca oleracea* L. and its antiacetylcholinesterase activity. *Natural Product Research*, *33*(18), 2583–2590.

Yang, Y., Yang, X., Zhang, X., Song, Z., Liu, F., Liang, Y., … & Guo, Y. (2019). Bioactive terpenoids from Euonymus verrucosus var. pauciflorus showing NO inhibitory activities. *Bioorganic Chemistry*, *87*, 447–456.

Yew, P.-N., Lim, Y.-Y., & Lee, W.-L. (2019). Tannic acid-rich porcupine bezoars induce apoptosis and cell cycle arrest in human colon cancer cells. *Pharmacognosy Magazine*, *15*, 523–531.

Zhang, Q. W., Lin, L. G., & Ye, W. C. (2018). Techniques for extraction and isolation of natural products: A comprehensive review. *Chinese Medicine*, *13*(1), 1–26.

6 Storage and Quality Degradation of Dried Herbs, Spices, and Medicinal Plants

Harshavardhan Dhulipalla and
Hari Kavya Kommineni
Vignan's Foundation for Science, Technology and Research

V. Archana
Vanavarayar Institute of Agriculture

Lavanya Devaraj and Irshaan Syed
Vignan's Foundation for Science, Technology and Research

CONTENTS

DOI: 10.1201/9781003269250-6

6.1 INTRODUCTION

Spices are one of the oldest food materials used extensively to date. These spices' usage depends on the colour, flavour, bioactive elements, vitamins, phenols, and carotenoids (Šojić et al., 2015). Spices, herbs, and medicinal plant's value is directly related to the percentage of the bioactive compound. Spices have become a significant part of exotic cuisines, condiments, and seasonings. Every bioactive compound has its response to various diseases. In terms of herbs and medicinal plants, they are often associated with defensive properties present in them. Over time, the bioactive compounds that showcase defensive properties tend to decrease. Proper storage is required to reduce the degradation of these materials. Oxygen, microbial contamination, moisture, and light have been the major factors for this quality degradation (King, 2006). The price of packaging materials has decreased because of their technical advancements, alongside plastic films' intrinsic characteristics make them ideal for packing food. The idea behind the active packaging technology is to maintain the micro-environment inside the package while allowing for interaction between the food and the material. It aims to increase the shelf life of the food, preserve the nutritional, sensory quality, and offer microbiological protection.

Storage plays a significant part in retaining various compounds. Many packaging approaches have turned up, including incorporating oxygen, scavenging chemicals, anti-microbial agents for microbiological safety, ethylene scavengers for oxygen, and moisture-sensitive foods. These incorporations are either made into the packaging material or between the commodity and material. Various approaches to packaging have been established to minimize food deterioration over time, and such materials are termed active packaging, intelligent packaging, and interactive packaging. Technological advances in the field of packaging raised the bar of preserving and keeping the commodity's quality with biofilms. These materials are considered sustainable materials that maintain product quality and the environment. The quality of herbs, medicinal plants, and spices tends to lower and reducing the product value. The materials were made available in the dried form, making them easier to utilize and enhancing their shelf life. Along with this, many factors are contributing to the degradation during storage.

Spices are available in various forms, such as fresh, dried, ground, frozen, pastes, extracts, crushed, and oils. Professionals extensively use fresh spices and herbs when preparing food due to their flavours, texture, and aroma. One thing to be considered while using fresh commodities is the reduced shelf life and stability. Hence, dried spices are made throughout the year at affordable costs but lack essential aromatic properties. It has been observed that fresh herbs and medicinal plants use a patented technology that provides fresh products with a refrigerated shelf life of 90 days (King, 2006).

The spices are ground or crushed to create spice extracts by applying extraction techniques such as solvent extraction, steam distillation, and other extraction techniques. Spice flavour-giving volatile and nonvolatile ingredients are concentrated versions employed to maintain flavour, colour, and aromatic consistency. Classification of spices along with flavourings is given in Table 6.1. Essential oils and other volatile substances give spices their characteristic scent. The aquaresins and oleoresins are composed of hydrophilic chemicals, resins, fixed oils, antioxidants, and gums that affect flavour or bite and considered nonvolatile. Oleoresins usually lack volatile components. Therefore, essential oils and oleoresins are required to generate a more comprehensive spice profile. The proper storage and packaging methods must be used to guarantee that spices and herbs preserve their flavour qualities for as long as feasible.

TABLE 6.1
Classification of Spices Along with Flavourings

Spice	Part of Plant	Flavouring Compound
Allspice	Fruit	Eugenol, β-caryphyllene
Anise	Seed, Fruit	(E)-anethole, methyl chavicol
Black Pepper	Berry, Fruit	Piperine, S-3 carene, β-caryophyllene
Caraway	Fruit, Seed, Root	d-Carvone, crone derivatives
Cardamom	Fruit, Seed	α-terpinyl acetate, 1–80-cineole, linalool
Cinnamon, cassia	Bark, Leaf	Cinnamaldehyde, eugenol
Chilli	Fruit	Capsaicin, dihydro capsacin
Clove	Flower bud	Eugenol, eugenyl acetate
Coriander	Leaf, Seed	d-Linalool, C10-C14-2-alkenals
Cumin	Seed	Cuminaldehyde, p-1,3-mentha-dienal
Dill	Fruit, Leaf, Top	d-Carvone
Fennel	Leaf, Twig, Fruit	(E)-anethole, fenchone
Ginger	Rhizome	Gingerol, Shogaol, neral, geranial
Mace	Aril	α-Pinene, sabinene, 1-terpenin-4-ol
Mustard	Leaf, Seed	Ally isothiocyanate
Nutmeg	Seed, Kernel	Sabinene, α-pinene, myristicin
Parsley	Lead, Seed, Root	Apiol
Saffron	Stigma of Flower	Safranol
Turmeric	Rhizome	Turmerone, Zingeberene, 1,8-cineole
Vanilla	Beans, Pods	Vanillin, p-OH-benzyl-methyl ether
Basil	Leaf	Methyl chavicol, linalool, methyl eugenol
Bay laurel	Leaf	1,8-Cineole
Marjoram	Flower, Leaf	e- and t-sabinene hydrates, terpinen-4-ol
Oregano	Leaf, Flower	Carvacrol, thymol
Origanum		Thymol, carvacrol
Rosemary	Terminal shoot, Leaf	Verbenone, 1–8-cineole, camphor, linanool
Sage, Clary	Terminal shoot, Leaf	Salvial-4 (14)-en-1-one, linalool
Sage, Dalmation	Leaf	Thujone, 1,8-cineole, camphor
Sage, Spanish	Leaf	e- and t-sabinylacetate, 1,8-cineole, camphor
Savoury	Floral, Leaf	Carvacrol
Tarragon	Leaf	Methyl chavicol, anethole
Thyme	Terminal shoot, Leaf	Thymol, carvacrol
Peppermint	Leaf, Terminal About	1-Menthol, menthone, menthfuran
Spearmint	Leaf	1-Carvone, carvone derivatives

6.2   PRESENT STORAGE PRACTICES USED FOR DRIED HERBS, SPICES, AND MEDICINAL PLANTS

Each spice, herb, and medicinal plant has its way of storage. The processing and storage aspects change with every spice. The methods followed are dependent according to the physical-chemical properties of the material, the volatility of the bioactive compound, their usage, and their structure.

6.2.1 Packaging Materials and Their Characteristics

The main goal of packaging is to maintain the product's quality and flavour until it is delivered to the consumer. When selecting a good packing material for flavoured food, a wide range of elements must be carefully considered. Essential elements and consumer approval criteria can be used to categorize the variables. The fundamental elements to be considered while selecting a packaging material are

- cost of the material,
- the rate of protection for the commodity from contamination,
- resistance towards impacts and injury,
- easier handling and storage practices,
- compelling interest in the interior surface, and
- overall features which relate to the overall performance of the packaging material.

6.2.1.1 Cardboard and Paper Cartons

The cardboard and paper cartons are referred to as the most convenient and cheapest form of packaging. The primary advantage of paper and cardboard cartons lies in folding into any form and providing space for advertising the product. The material can be made in a water-resistant form by applying a wax coating on the exterior. A coat of polyethylene gives sealability and an additional layer of protection. They are often avoided for grounded spices as they are prone to significant changes. Hence, an internal packaging material made out of polyethylene is required.

In addition, a wide range of paper-based packaging materials is replacing the current materials. These materials include

- kraft paper
- bleached paper
- wrapping tissue paper
- grease-proof paper
- glassine paper
- vegetable parchment paper
- waxed paper
- sulphite paper
- paperboard
- paper bags
- composite cans
- newspaper as packaging material
- fibre drums
- folding cartons
- corrugated fibreboard
- multi-wall paper sacks
- rigid boxes
- moulded pulp packaging
- paperboard-based liquid packaging
- paper labels and adhesives

Using paper-based materials in primary and secondary food packaging is growing daily. However, ignoring their detrimental effects on human health and environment is impossible. Functional and processing additives such as phthalates, mineral oil, and others transferred onto food items from recycled paper, paperboard, and recycled paper do not meet the requirements for direct-touch food-packaging applications (Deshwal et al., 2019).

6.2.1.2   Aluminium Foil

Aluminium foil is an ideal packaging material for grounded spices that require protection from light due to its nature of being non-transparent. It can potentially resist gas transmission, which protects the flavours of the spice. The material tends to tear down quickly and can be avoided using a lamination of paper. The potentiality of heat sealability can be obtained by the application of a layer of polyethylene to the material.

The conjugates of metalized films and laminates made of plastic or paper provide the packed formulation significant flexibility (Singh et al., 2011). Laminated aluminium is frequently used for expensive packaging delicacies such as dry soups, herbs, and spices. A metalized film that can withstand moisture, grease, and air has been created cost-effectively and is mainly utilized for snack-packaging. Regularly packaged in metal containers, the dietary supplements maintain their shelf life for years without losing any nutritional content (Verma et al., 2021).

6.2.1.3   Glass

Glass can be made in any shape, form, or size. Along with that, the surface of the glass can be treated with various other compounds to increase their strength and rigidity. Glass can be coated with a range of colours, making it feasible for storing volatile and light-sensitive compounds.

6.2.1.4   Flexible Films

A film can be mixed with other entities to get the desired effect because a single film cannot meet all functional criteria. Co-extrusion, coating, or lamination can accomplish this.

6.2.1.5   Single Films

A wide range of single films has been made available in the market, which are utilized according to the type of material used and its requirement.

6.2.1.5.1   Cellulose Films

Cellulose appears plain, odourless, glossy, tasteless, and transparent and degrades within about 100 days. Although it shreds readily, it is resilient and resistant to punctures. It is perfect for twist wrapping because of its low slip, dead-folding, and static build-up resistance. The film's size and permeability change with humidity and cannot be thermally sealed. Fresh bread and various varieties of confectionery are among the goods that employ it that requires entirely adequate moisture or gas barrier.

Polyethylene terephthalate (PET) is a glossy and transparent material that is exceptionally robust and has good gas and moisture characteristics. It shrinks relatively little with changes in temperature and humidity and is elastic at temperatures between –70°C and 135°C.

Some tubs and trays employ co-polymers such as low-density polyethylene (LDPE). It shrinks when heated and is chemically inert, sealable, and odourless. Although it has high gas permeability, an effective moisture barrier is sensitive to oils and slightly resistant to smell. For safe stacking, low slip qualities can be used; on the other hand, high slip properties make it simple to stack them in an external box. Since it is the most affordable of most films, it is frequently utilized.

High-density polyethylene (HDPE) has a lower penetrable nature to moisture and gases than low-density polyethylene and is more potent, brittle, less flexible, and thicker. High rip strength, resistance to penetration, tensile strength, sealability, and penetration resistance are all characteristics of sacks manufactured of 0.03–0.15 mm HDPE. They are used in place of multi-wall paper bags as drinking containers since these materials are chemically resistant and waterproof. Uncoated ethylene vinyl acetate, polyvinylidene chloride (PVdC), and polystyrene are other forms of film architecture (EA).

6.2.1.5.2 Coated Films

Additional aluminium or polymers are provided to coated films to enhance their barrier qualities or provide films with the potential to heat seal. A thin layer of aluminium produces a highly effective barrier against moisture, oils, gases, odours, and light. The metallized film has more flexibility and cost efficiency than foil laminates, which have comparable barrier capabilities. Although metalized polypropylene is more extensively used than metalized polyester owing to economic reasons, the latter offers greater barrier qualities.

6.2.1.5.3 Laminated Films

A package's aesthetic, barrier qualities, and mechanical strength are all enhanced by laminating two or more films. For non-respiring items, laminates often include nylon-LDPE, nylon-EVOH-LDPE, and nylon PVdC-LDPE. Nylon offers strength to the pack, EVOH or PVdC gives it suitable moisture and gas barrier qualities, and LDPE gives it the ability to be heat sealed. In addition, PVC and LDPE are utilized to make typical MAP goods.

6.2.1.5.4 Co-extruded Films

Co-extrusion is a process of creating a single film with the extrusion of two or more polymer layers. They have significant other advantages when compared with existing packaging materials. They are made at a lesser cost and possess very high barrier qualities, close to multi-layer laminates. Their thickness makes them simpler to utilize in forming and filling equipment. They are more attenuated than laminates and are closer to mono-layer films. Their properties make sure that the layers do not separate easily.

The significant compounds used in the preparation of this material are

- olefins: a low-density polyethylene and high-density polypropylene;
- styrene: it is a combination of acrylonitrile-butadiene-styrene and polystyrene; and
- a part of polyvinyl chloride polymers.

6.2.1.5.5 Biodegradable and Edible Films

Growing environmental consciousness has caused a paradigm change, leading people to search for degradable packaging films and procedures; hence, the environmentally friendly concept of degradation benefits from both users and eco-friendly characteristics. The significant materials are primarily obtained from marine processing and agricultural feedstock, which allows for resource conservation with an emphasis on a safe and eco-friendly environment.

Biopolymers obtained from various resources and agricultural feedstock can produce suitable packaging materials upon blending and processing. Utilizing them with other components, such as plasticizers and additives, can better express their functioning. The advantages of such bio-polymeric packaging material are as follows:

- It can be used as a disposable packaging material.
- It has the potential to be used every day.
- It can be added with a lamination coating.
- They can be utilized for agricultural uses.

Two different biomolecules, lipids and hydrocolloids, are combined to create biodegradable packaging films or composites. They are structurally weak and functionally unremarkable on their own. Lipids are excellent moisture barriers that are added to hydrocolloids to compensate for their hydrophilicity, making them poor. These and other components are combined in various ratios to create composite films and mechanical characteristics and determine the film's barrier (aroma compounds, carbon dioxide, oxygen, and water) characteristics.

6.2.1.6 Modified Atmosphere Packaging

It is implied by the phrases "controlled atmosphere" and "modified atmosphere" that gases are added to or subtracted from depository spaces, packages, or containers held for transportation to control the concentrations of gases like nitrogen, carbon dioxide, ethylene, and oxygen, among others. The modified environment is more frequently employed to preserve food goods and avoid unfavourable alterations to their wholesomeness, safety, sensory qualities, and nutritional value. The objectives mentioned above are achieved based on a few principles:

- Inhibiting the microbial growth.
- Reducing the undesirable physical and chemical changes.
- Prevention of cross-contamination.

6.2.1.7 Active Packaging Technologies

The four basic categories of active packaging are carbon dioxide release and scavenging, anti-bacterial, and scavenging of oxygen. Three main factors determine the sort of active packaging that should be used. In this packaging technique, food requirement is prioritized, and subsequently, the container and the demand for the active ingredient.

To retain the sensory and safety characteristics and preserve the quality of packed food, the active packaging technique is used as it can alter the state of the packaging and maintain such features during the storage duration. If the product is misused, the state of the food may degrade, and the customer may not be aware of this till the food package is opened. By lowering undesired needs, active packaging materials are intended to actively preserve or improve the condition of the food (Nura, 2018).

6.2.1.8 Intelligent Packaging

The term "intelligent packaging," also known as "smart packaging," refers to packaging that senses specific characteristics of the food it contains or the environment in which it is stored and alerts the consumer, retailer, and producer to the status of these characteristics and conditions during transport and storage. Nanosensors can detect the presence of gases, odours, chemical pollutants, diseases, and even environmental changes when used in food packaging, which can answer this problem (Nura, 2018).

6.2.2 STORAGE OF SPICES

6.2.2.1 Coriander

The best way to store ground coriander is in aluminium foil pouches, which may keep it fresh for up to 6 months. Large volumes of powder can be stored in Jute bags lined with polyethene. Bags made of paper, polyethene, or cotton should not be used for storage. Given their vast exposed surface area, ground spices must be appropriately shielded from air to slow down the rate at which their flavour degrades. Coriander seed may be kept for 6 months in cotton or polythene bags with no flavour loss.

It has been found that a 20%–25% decrease in volatile oils after a year. Hence, it is required to utilize flexible laminated foils and plastic films, which have higher physical and chemical qualities to protect the product against volatile oil loss, fat seepage, and moisture intrusion during storage. To avoid browning, flavour loss, and moisture intrusion, the seeds are recommended to preserve in a cold, dark, and dry environment. Coriander seed is observed to maintain flavour and colour for 6–9 months when stored properly.

6.2.2.2 Garlic

In big bulk bins or open mesh bags, garlic picked for dehydration is transported to a dehydration facility. The garlic is separated using rubber-covered rollers that apply the required amount of pressure to obtain the cloves without crushing them, later divided into individual cloves. Aspiration and screening are used to get rid of the loose paper shell. The root stabs are floated off while the cloves are rinsed in a flood washer. The same technique is used to slice and dry onions and dehydrate garlic. Commercial garlic drying removes roughly 6.5% of the moisture. Commercially available forms of dehydrated garlic include garlic powder, granules, slices, chops, and mince.

Garlic is dried at temperatures between 50°C and 70°C for 5–8 hours to produce garlic powder, which results in volatile flavour losses of up to 30%–35%. The unwanted yellowish-brown powder is produced by non-enzymatic browning processes. Due to the powder's high hygroscopicity and requirement for packing with little water vapour transport, clumping is another problem.

6.2.2.3   Turmeric

When dried and crushed, this plant's rhizome produces a flavorful yellow powder used for generations as a natural colourant and flavouring in foods, clothing, and cosmetics. Typically, turmeric drying is done by artificial means or the sun and then milled into a powder primarily used as a colouring.

6.2.2.4   Cinnamon

The inner and outer bark of cinnamon is stripped away during preparation. The inner bark naturally folds into quills, smaller pieces, and quills are filled to form a nearly solid cylinder linked to extending their length. Smooth, uniform quills of a high calibre are in golden-brown colour. Smaller bark fragments and quills are sold individually, and a part is crushed or chopped for distillation or local sale. The best bark should be wrapped up in new corrugated boxes or sacking, and cinnamon quills should be kept in bags. For shipping, quills are compressed into cylindric bales of 100–107 cm and weighing 45–50 kg. The bark is packed into distinct containers depending on distinctively acknowledged norms.

6.2.2.5   Cumin

Cumin is one of the spices that can now be grounded using a grinder, which exposes the spices to high heat. Cumin is more volatile when grounded at colder temperatures, enhancing particle quality and sensory properties. An optimum storage environment includes aluminium and polyethylene pouches, with a relative humidity of 70% and temperature of 37°C.

6.2.2.6   Nutmeg

The aril (mace) is extracted from the fruit, and the aril surrounding the fruit is split apart. Nuts are shelled after drying to produce the spice nutmeg. Nutmeg trees produce three primary products: mace and nutmeg, which are used as spices and flavourings; nutmeg, mace oils, oleoresins, leaf oil, and other derivatives. The most significant home products are mace and nutmeg, although oleoresins and oils are increasingly used in industrial settings.

Mace and nutmeg should be grounded only when necessary since the organoleptic properties quickly degrade due to the loss of volatile oils. Improperly kept nutmeg oil may also experience severe compositional changes when exposed to a warm environment. Powders and oils that are not covered might absorb offensive odours. Until needed, oleoresins, oils, and powders should be kept in complete, airtight containers made preferably of opaque glass. In importing nations, nutmeg is offered whole rather than further processed to a spice powder with a particular mesh size. Nutmeg spice can be marketed in various packaging but is typically provided in high-barrier plastic packing film or glass to preserve the product's quality. Whole nutmegs are packaged in bags.

6.2.2.7   Mustard

The most valuable commodity, mustard, produces a seed with a 30% oil content. The seed must not be warmed during drying or storage as this might result in rancidity and quality loss. When it is received for storage, the moisture level of the seed ranges between 10% and 25% for standing crops and between 10% and 15% for windrowed. Due to their hard exterior shell, dry and clean seeds store well, but proper packing stops the round seeds from moving around the material.

The mustard seed is kept in bulk or in sacks. The mustard meal is made by grinding dry, entire seeds, and it should be stored in a cold, dark place, either fresh or sealed. This product is more frequently employed in the pharmaceutical sector than as a food-based product.

6.2.2.8 Pepper

One of the most common spices in the world is pepper. Pepper is made in a variety of forms. After harvesting, the pepper berries are separated from the spike by rubbing between the hands or stomping with the foot. After being immersed in hot water for 1 minute, the green pepper is swiftly dried in the sun for 5–7 days to achieve a moisture content of 10%–11%. The dried pepper is cleaned by removing the husks, stems, and pinheads.

The result made from completely ripe berries is white pepper. The seeds are collected and heaped to ferment while soaked in water for 5–7 days. Then, the outer coating and pulp of the seed are removed. White pepper has a smooth exterior and a yellowish-grey colour. In addition, black pepper is made into it by mechanically grinding off the exterior bits.

6.2.2.9 Ginger

After 8 weeks (23%) and a further reduction (37%), gingerol in powdered ginger stored in glass jars at 4°C significantly decreased. In contrast, the gingerol showed a slight drop when stored in ambient storage at 23°C after 8 weeks, the drop was 30%, and after 16 weeks, it was 37%.

6.2.3 LATEST TRENDS IN SPICE STORAGE UTILIZING IRRADIATION

6.2.3.1 Irradiation

During "irradiation," food is subjected to ionizing radiation, such as gamma rays from the radioisotope 60Co, produced by mechanical sources. It is a powerful technique for eliminating foodborne germs without affecting their physicochemical or sensory characteristics. The DNA of organisms may be damaged effectively by gamma radiation, which prevents organisms from proliferating and growing in food. Ionizing radiation causes energy absorbed by food products, which causes interactions between atoms and molecules that render microbes inactive (Rahman et al., 2021). A schematic diagram for irradiation treatment is given in Figure 6.1.

Consuming food that is fresh, little processed, or ready to eat is becoming more popular. The findings linking fruits and vegetables to several epidemics are a cause for alarm on a global scale. Irradiation is a cold procedure that offers a potentially effective way to combat the contamination of fresh vegetables with human bacterial infections and parasites. The spice storage with response to irradiation dosage is given in Table 6.2.

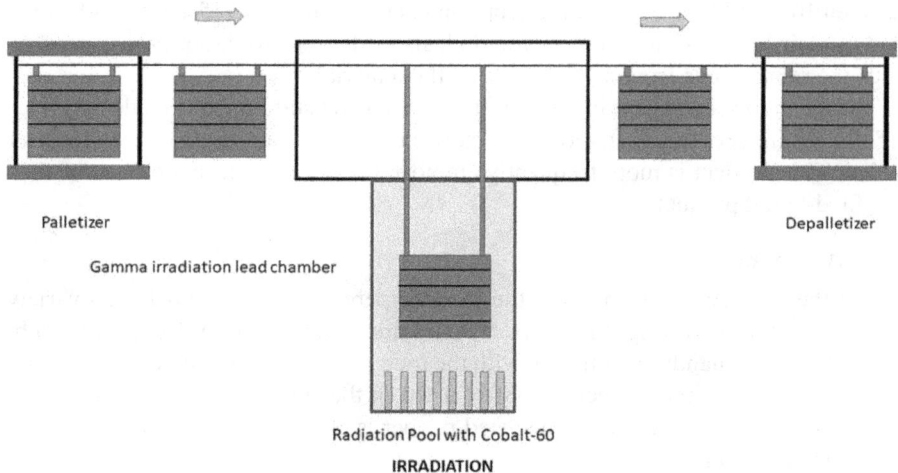

FIGURE 6.1   Schematic diagram for irradiation treatment.

TABLE 6.2
Spice Storage with Response to the Irradiation Dosage

Spice	Irradiation Dosage (kGy)	Storage
Paprika	6–9	Nine months
Turmeric	5	Eight months
Cumin	6	Three months
Coriander	7	Six months
Garlic	4	Ten months
Black Pepper	4–6	Six months

6.2.3.2   Paprika

The Himalayan paprika was analysed for 9 months to understand the various changes undergone in the product. Total phenolics, carotenoids, and capsaicinoids were retained in 9 kGy-irradiated samples at percentages of 88.7%, 72.11%, and 78.0%, respectively, vs 79.5%, 61.54%, and 69.5% in control samples (Ayob et al., 2021). After up to 9 months of ambient storage, gamma radiation therapy at doses between 6 and 9 kGy was highly efficient in maintaining the bacteria burden below sensing limits. It might be suggested that a gamma irradiation dosage of 6 kGy is the minimal dose necessary to maintain sanitization and safety for the entire ambient storage period of 9 months without violating Himalayan paprika quality criteria.

In a research study relating to the comparison of storage effects of gamma radiation on dehydrated paprika and sundried paprika, dihydrocapsaicin and capsaicin showed a relation between drying methodology and storage stability. Out of which, the change in pungency is relatively lower in dehydrated paprika, whereas a notable difference has been observed in sundried paprika during storage after 2 months (Topuz and Ozdemir, 2004).

6.2.3.3 Turmeric

Several radiation dosages were administered to the turmeric samples (5, 10, and 12.5 kGy). As the dosage gradually increases, all of these spices' counts simultaneously fall (Munasiri et al., 1987). All bacteria were successfully eliminated by a 10 kGy dosage, leaving zero counts. All molds may be eliminated with a 5 kGy dosage (Rahman et al., 2021). The bacterial and fungal counts of both non-irradiated and irradiated samples somewhat decreased during storage. In the case of radioactive samples, the decline coincided with an increase in dosage. During storage, the stability of curcumin in ground turmeric was decent (Chatterjee et al., 1998).

Spices must have less than 10^2 cfu/g of bacteria, as mandated by the importing nations, and a dosage of 5 kGy might accomplish this. It was also discovered that this dose eliminated the fungus in ground spices. All of these spices might be sufficiently disinfected with a dosage of 10 kGy. The Joint FAO guidelines suggest an average dosage of 10 kGy.

6.2.3.4 Cumin

The non-irradiated and irradiated cumin's DPPH scavenging activity varies. The radical-scavenging ability of the cumin ethanolic extract assessed by DPPH showed a slight increase after irradiation at 1, 3, 5, and 10 kGy. However, this increase was not statistically significant compared to the non-irradiated cumin. However, the 10 kGy-irradiated cumin showed a non-significant decrease compared to the 5 and 3 kGy-irradiated cumin. The abundance of amino acids and sugars in cumin may cause the gradually rising DPPH scavenging activity.

As a result, the natural antioxidants in cumin were maintained after gamma irradiation at 1, 3, and 10 kGy, which is essential for the quality of spices. The DPPH scavenging activity was maintained well and even increased slightly. The irradiated cumin samples of 5 and 10 kGy showed a considerable rise in the antioxidant index. The total amount of irradiation cumin retained its reduction power compared to unirradiated cumin. All irradiation cumin showed little maintenance and increased its overall polyphenolic content (Kim et al., 2009).

6.2.3.5 Coriander

In two trials, coriander leaves packed in polyethylene sachets of 25 g were collected and stored in melting ice and subjected to doses of 1, 2, and 3 kGy of gamma radiation at a dosage rate of 0.02 kGy/min. The sachets were subjected to a single dosage of 1 kGy chlorine treatment in different studies. The smallest dose of 1 kGy reduced bacteria by three log cycles, yeast, and mold by one log, and coliform by 43 cfu/g. After 1 week of storage, an increase in total mold (TMC) and total bacterial (TBC) count is seen in unirradiated samples. However, the microbial burden in untreated samples did not significantly increase after additional 1-week storage.

On the other hand, the survivors (bacteria and molds) grew slowly in irradiated samples over 2 weeks of storage. Coliforms could be effectively reduced and eliminated with low-dose irradiation (1 kGy), and there was no sign that they would reappear during storage. The plant tissues can tolerate gamma irradiation when the doses are less than 1 kGy. Since these pathogens are usually present in leaves, the minimal dosage (1 kGy) would be sufficient to remove their burden (Kamat et al., 2003).

6.2.3.6 Garlic

The samples of garlic that have been procured were treated at Co-60 with a dosage of 0.03 kGy at a rate of 0.41 Gy/s. Non-irradiated garlic bulbs biologically lost more weight than irradiated ones after 300 days of storage. The weight losses in the irradiated and non-irradiated garlic bulbs following the storage time were 55.0% and 24.0% of the original weight, respectively. Studies show that the marketable garlic bulbs on plastic trays of irradiated garlic bulbs were more significant than those on non-irradiated garlic bulbs after 210 days of storage. The sprouting and rotting of non-irradiated bulbs led to their disposal, whereas the rotting of irradiated bulbs was the only reason. Similar outcomes were attained with the piled-up garlic bulbs.

Inner sprouting was missing in the cloves of marketable irradiated garlic bulbs during storage while being present in 50% of the non-irradiated garlic bulbs at the start of the testing period (120 days following harvest). 25% and 100% of the cloves had radiation-induced darkening (small spots) in the growth centre after 150 and 300 days of storage. According to studies, garlic extract typically maintains its stability at 55°C for up to 36 hours. Under these circumstances, the anti-fungal effect is only stable for up to 8 hours (Arora and Kaur, 1999).

6.2.3.7 Black Pepper

Based on the prior research, there are several suggestions for the recommended dose of irradiation to accomplish either a decrease in the microbial load or its total eradication in black pepper. To inhibit *Salmonella typhimurium* and *Escherichia coli*, a study conducted in 2014 recommended a dosage of dried red pepper of 5 kGy (Song et al., 2014). Another paper from 2015 suggested a 6 kGy to lower the overall aerobic microbe count while maintaining the red colour and pungency standards for red pepper powder (Jung et al., 2015). In 2012, a research revealed that pepper could be successfully decontaminated from the fungal population at a dosage of 6 kGy (Iqbal et al., 2012).

6.3 FACTORS AFFECTING QUALITY DEGRADATION DURING STORAGE

Most herbs are traded in dry form since a product's quality will deteriorate over time if it contains much water. The biological characteristics of the plants, the drying method, and the volatile content predominantly impact how the volatiles. To identify the most appropriate packaging material, it is required to understand the various factors triggering the product's deterioration.

   i. Light Sensitivity: Herbs containing chlorophyll and other carotenoids are susceptible to deterioration resulting in loss of colour and disturbances in the bioactive compound present in the material.

   ii. Moisture and Oxidation Sensitivity: With decreasing spice particle size, the product has exposed surface area and susceptibility to moisture penetration increases. Similarly, oxidation takes place. An increase in moisture content results in a spike in potential microbial risk. To prevent oxidation, the product is meant to avoid high-temperature storage; specific and optimized gas composition needs to be flushed into the packaging material with either a modified atmosphere or controlled atmosphere conditions.

iii. Flavor Sensitivity: Spices and herbs are given maximum priority due to essential oils, but they tend to deteriorate once harvested. Likewise, the gradual reduction in flavour is also seen.

iv. Grinding: After harvesting, the spices are reduced in their particle size, making them much more vulnerable to oxidation, flavour loss, and moisture increases. While grinding, the spices get exposed to high temperatures, giving rise to lower concentrations of bioactive compounds. To inhibit these damages, the temperatures, humidity, and oxidation reactions need to be controlled. The current method of grinding spices has several inherent drawbacks, including significant heat generation, low efficiency, and volatile oil loss. For components of plant origin, such as spices with high fibre and fat content and heat sensitivity, this grinding method is not preferred. Reduce or maintain the mill temperature as low as feasible (which can be relatively below the boiling point of the particular compound in the spice oil) to partially compensate for volatile oil loss. While whole spices have a one-and-a-half-year shelf life, ground spices have a 3–4-month shelf life when kept at a cool temperature.

Apart from the abovementioned factors, a few main factors cause food deterioration during storage.

a. The climatic influences trigger the physical and chemical changes in the spices and herbs (moisture, oxygen, UV light, and temperature changes).

b. Cross-contamination occurs due to microbes, insects, or other foreign materials.

c. The compression and impact factors often seen during storage and transportation are also factors.

d. Adulteration and tampering with the product also affect the quality of spices and herbs.

6.4 IMPROVED/RECOMMENDED STORAGE PRACTICES FOR SAFE STORAGE (DRIED HERBS, SPICES, AND MEDICINAL PLANTS)

The purpose of storing and conserving medicinal plants is to prevent their quality from degrading by maintaining both their quantitative and qualitative characteristics after drying, creating ideal conditions for relative humidity and temperature and avoiding the attack of insects, fungi, and microorganisms during the storage period. Reduced metabolic activity during storage makes medicinal plants less prone to degradation. This can be accomplished by a cooling system that holds the medicinal plants, using a modified environment, or lowering the product's water content to safe limits. Over time, the product starts to degrade due to high water concentration. Hence, most herbs are sold in dry form. There is a considerable loss in volatiles when dill and parsley are oven-dried or frozen.

6.4.1 SPICES

At the very least, they must be kept cool and dry. They should never be close to a smokehouse, a kitchen range, or another intense heat source. They should also never be left near a dishwasher, a washroom, or any other area where there is a chance that the containers will get wet. The containers in large-scale storage should always be palletized and kept away from exterior walls. They must be placed at 68°F (20°C) and 60% relative humidity.

As per the research studies, cold storage (between 32°F and 45°F) is strongly recommended, especially for spices and capsicums (such as paprika, red pepper, etc.), where volatile oils or distinctive scents are crucial quality characteristics to be evaluated. Paprika will lose 1% of its colour every 10 days at a temperature of 70°F–80°F. Losses happen significantly more quickly when the temperature increases. However, paprika's colour loss under cold storage is decreased to 2% every 10 days, meaning it can be kept in good condition for up to 6 months. Infection and rancidity can also be retarded by using cold storage in fixed oils (such things as sesame and poppy seeds) (such items as sesame and poppy seeds).

Paprika, parsley flakes, chives, and other light-sensitive ingredients must be shielded from direct sunshine and fluorescent lighting.

6.4.2 DRIED HERBS

Herbs can be preserved by using freeze-drying techniques to remove moisture. The best and most efficient method of dehydration is freezing drying. The reason for this is that the herbs' quality is preserved, which means that compared to other drying techniques, the scent, shape, and form of the material retain superior quality. Herbs are freeze-dried to maintain their quality and extend their shelf life. Freezing drying is one of the techniques for drying food that comes closest to producing the desired outcome since it keeps the structure and flavour of the herbs.

Freeze-dried herbs have a strong fragrance and are nutrient-dense. Today, Germany, Austria, France, Italy, the United Kingdom, and Switzerland are the countries that utilize freeze-dried herbs most frequently. Given the significance of freeze-dried herbs at the time, in addition to common herbs like mint and tea, pure compounds isolated from plants are also utilized, following laboratory testing, to treat ailments. Herb consumption may aid in treating and managing diabetes, cancer, and heart disease. In comparison, additional studies are required to validate their therapeutic and health benefits.

6.4.3 MEDICINAL PLANTS

To preserve and store medicinal plants, appropriate relative humidity and temperature conditions must be maintained, as well as protection from insects, fungi, and microorganisms during storage. This is done by maintaining the quantitative and qualitative aspects after drying. Reduced metabolic activity during storage should make medicinal plants less prone to degradation. This can be done by cooling the system where medicinal plants are stored, utilizing a modified environment, or lowering the product's water content to an appropriate level.

One year is the suggested shelf life for medicinal herbs. However, if the maker provides stability test results that show that the product's qualities will remain intact for the requested duration, and then a longer shelf life may be acceptable. Before being employed as a raw material to create various goods, medicinal plants are stored for extended periods (Sourestani et al., 2014). However, incorrect storage could alter the chemical, physical, and biological systems (Mayuoni-Kirshinbaum et al., 2013).

Depending on how long they are held, chemical and active chemical concentrations can change. The essential oil content of *Thymus daenensis* Celak shoots was examined by Rowshan et al. (2013) at various storage conditions, including the refrigerator (–4°C), freezer (–20°C), and room temperatures (25°C). No adverse effects on the essential oil's quality were observed at room temperature storage and its constituents, such as thymol and carvacrol.

The kind of packaging to use while preserving medicinal plants is crucial. The most common containers used nowadays for storing plant leaves are double kraft paper bags, polyethylene bags, paper bags, kraft paper bags, and double kraft paper bags with an inner layer of non-toxic polyethylene.

After a year of storage, Costa et al. (2009) noticed a considerable decline in the actual oil output of dried samples of whole and crushed *Ocimum selloi* leaves packaged in polypropylene bags. The impact of storage duration on the chemical and yield makeup of this plant's essential oil was also investigated. Nevertheless, compared to pulverized leaves, the yield from undamaged leaves was much higher.

6.5 PRECAUTIONS NEED TO BE TAKEN TO AVOID MICROBIAL GROWTH, AFLATOXIN DEVELOPMENT, AND QUALITY DEGRADATION IN DRIED HERBS, SPICES, AND MEDICINAL PLANTS

Up to 25% of the world's food crops may contain mycotoxins, according to the Food and Agricultural Organization (FAO) of the United Nations (WHO, 1999). Due to the damages they inflict, their effects on animal and human health, and the national economic effects of all of these factors, mycotoxins are significant on a global scale (Bhat et al., 2010). Unfortunately, many spices are highly sensitive to toxic fungal strains and are likely to be contaminated with aflatoxins (AFs). Most naturally occurring AFs are produced by the aspergilli *Aspergillus parasiticus* and *A. flavus*. Rarely generating AFs are the species *A. pseudotamari, A. bombysis, A. nomius*, and *A. ochraceoroseus*. In underdeveloped nations (such as India, Sri Lanka, and Pakistan), where there is a lack of infrastructure to protect food items from fungal infection, there are issues with AFs. Aspergillus growth and proliferation, which creates a risk of AFs contamination, may be facilitated by non-compliance with good hygiene practices, agricultural practices, manufacturing practices, storage practices, and shipping methods. Pollutants in herbs and spices can be reduced or eliminated through the use of good manufacturing practices (GMP), ISO 9000 quality management systems, hazard analysis and critical control points (HACCP), and good agricultural practices (GAP) (George, 2006; Steinhart et al., 1995). Numerous investigations found significant microbial contamination in herbs, spices, and medical plants. Inaccurate growing, harvesting, storage, or processing can increase the aerobic bacteria and fungi naturally present in plant material.

It has been demonstrated that herbal substances can contaminate pathogenic organisms such as *Enterobacter, Enterococcus, Clostridium, Pseudomonas, Shigella,* and *Streptococcus.* Setting limits for microbiological contamination is crucial, and the *European Pharmacopoeia* now provides non-obligatory recommendations on appropriate levels (Barnes et al., 2007; Nafiu et al., 2017). Better control and precautions should be used in all facets of these products production, processing, and use to prevent microbial growth and aflatoxin development, which may ultimately lead to quality degradation. Contaminated spices, medicinal plants, and herbs bring on these hazards.

Although reducing fungal growth is crucial to lowering human exposure, it is difficult to eliminate or prevent AFs contamination during pre- and post-harvest procedures. Numerous physical, pharmacological, and biological techniques have been proposed to detoxify or remove AFs. There are two basic categories for AFs treatment strategies. Both of these strategies lower the levels of AFs in food commodities: the first involves using excellent agricultural practices to decrease AFs' incidence throughout the growing cycle, and the second entails limiting the accelerated toxin accumulation in the post-harvest supply chain. Farmers must utilize crop varieties that are resilient to the local growing environment, particularly drought, insects, and pests, as well as those that exhibit resistance to fungus contamination controlling humidity after harvest is essential for lowering the risk of AFs infection. Plant health can be improved by irrigation and fungicides to help them resist the fungus that causes AFs. Another way to manage the moisture in spices before storage is to use a solar drier. Furthermore, new detoxification technologies, such as gamma irradiation, microwave, ultraviolet, electrolyzed water, ozone, pulsed light, cold plasma, and pulsed light in combination with chemical, physical, biological, or genetic engineering techniques, may be applied for the purpose of detoxification. In addition, GAP, advanced agricultural technologies, good storage techniques, and suitable manufacturing processes can help lessen the possibility of contamination of AFs.

6.5.1 HERBS AND SPICES STERILIZATION

When spices and herbs are collected, they frequently have a large number of microorganisms present. Numerous factors, including poor drying, contact with infected surfaces, improper packing, and re-absorption of moisture during storage, contribute to the growth of the microbial population. As a result, it is typical to discover spices with plate counts of 10 million or more colonies per gram, such as black pepper. In addition to shortening shelf life, a high microbial load increases the possibility that hazardous species will create phytotoxins. As a result, microbiological treatment or sterilization of herbs and spices is necessary. Most food sterilization methods including heat, steam, chemicals, lyophilization, desiccation, dehydration, low temperature, acidity alteration, chemical preservative application, and irradiation are reviewed to eliminate molds and other germs in herbs and spices (Plusquellec, 1995). Using chemical or steam fumigation for dried or ground herbs and spices seems preferable.

6.5.2 PHYSICAL METHODS

6.5.2.1 Hydrostatic Pressure

They can be physically decontaminated by exposing them to high hydrostatic pressure, typically between 300 and 1,000 MPa. There are 99.9% fewer germs in the colon rod, even at 100 MPa of pressure (*Escherichia coli*). According to hygienic regulations, these bacteria (*Staphylococcus aureus* and *Salmonella*), which are some strains that might cause harmful poisoning, are not authorized in products meant for human consumption. *Candida albicans*, a kind of yeast from the yeast white litter group, and *Saccharomyces cerevisiae* perish after 5 minutes when placed under hydrostatic pressure of 500 MPa. Although it dramatically lowers the number of essential oils in plants exposed to mold, this approach is nevertheless quite effective against it. The effectiveness of decontamination when carbon dioxide is used under high pressure is relatively low. The spores of mold and germs are not sufficiently destroyed. In addition, as the moisture level of the plant material declines, so does its efficacy. Further, carbon dioxide alters the chemical makeup of herbs and contributes to the extensive loss of aromatic oils.

6.5.2.2 Microwave Treatment

Microwaves are electromagnetic waves with wavelengths between 0.0001 and 1 m and are efficient against bacterial and fungal vegetative forms. Mold spores, Gram-positive bacteria from the genus *Clostridium*, *hay bacillus* (*Bacillus subtilis*) spores, and other bacterial species are resistant to their effects. As a result, their chemical composition is altered, and their aromatic oil concentration is significantly reduced.

6.5.2.3 Irradiation

Although the World Health Organization has deemed ionizing radiation safe, consumers disagree. While 10 kGy is considered the maximum acceptable dose, viruses and bacteria spores can sometimes only be successfully eliminated with doses up to five times greater. The herbs cleaned up using this approach experience undesirable chemical and aesthetic changes. However, 5–6 kGy are sufficient for various plant seasonings to ensure that their microbe content complies with the guidelines. It is crucial to stress that radiation is appropriate for heat-sensitive raw materials because decontamination is only incredibly infrequently linked to a rise in temperature. It is difficult to mention a significant reduction in germs in this situation because the application of infrared radiation only affects the surface of plants.

6.5.2.4 Steam Sterilization

Many herbal raw materials can be processed using steam, but seasonings are the ones that employ it most frequently. Typically, the steam's temperature ranges from 100°C to 200°C. The plant material is then rapidly cooled after being heated air-dried. As a result, there is a considerable decrease in microorganisms, particularly those of the *Enterobacteriaceae* family, specific granulomas, hay bacillus bacterium spores, *Clostridium* bacteria, and yeast and mold. A sterilizer is a piece of standard equipment for disinfection using the steam method. Various plant raw materials can be sterilized using modern sterilizers, including whole herbs, ground spices (pepper,

TABLE 6.3

Advantages and Disadvantages of Various Physical Preservation Methods

Type of Method	Advantage	Disadvantage
High Hydrostatic Pressure	High biocide effectiveness against fungus and bacteria.	Essential oils are lost from the primary material.
Ionizing Radiation	The WHO highlights a consumer-safe approach.	It results in detrimental chemical and aesthetic changes in the source material. Efficacy varies with the dose.
Steam	Acquiring excellent microbiological measurements of sterile raw materials with little essential oil losses. A high throughput (about 1,000 kg/h) with little raw material losses. Inexpensive techniques without complex generators or no complicated chemicals are required, just a sterilizer.	The process results in highly noticeable colour change for plants with carotenoid and chlorophyll dyes. Drastically depletes the content by adding countless substances with biological activity. Ineffective for decontaminating raw powder. Materials—caking is the result.

nutmeg, cloves, etc.), and materials that have been sliced into smaller pieces (sawn herbs, fixed type fraction, cut herbs, grain yield of herbs, marjoram, allspice, coarse pepper, etc.). The Polish company sterilizes the initial product in the Scorpion sterilizer at around 90°C. (Rozdra new, Poland). The raw material is moved from the feeding hopper to the sterilizing chamber and heated to 130°C by dry steam exposure. The duration of this step is determined by the form of the raw material and, rarely, the type of plant being used. The raw material is moved into the heated air-pumped drying chamber from the pre-cooling room. The herbs are bundled into little or large bags using a receiving feeder.

The advantage of this technology is its potential to produce sterilized raw materials with highly favourable microbiological parameters while minimizing aromatic oil loss, raw material loss, excellent efficiency (on the order of 1,000 kg/h), and, ultimately, reasonably affordable sterilization costs. Currently, steam is the most popular method for sterilizing herbs, and contemporary approaches in this area ensure relatively good microbiological purity while being environmental-friendly. Advantages and disadvantages of various physical preservation methods are given in Table 6.3.

6.5.3 CHEMICAL METHODS

Chemical preservatives are chemical additives that are safe for human consumption and are applied to foods expressly to stop them from spoiling or decomposing. Chemical preservatives are primarily used in foods to prevent the growth and activity of bacteria that cause deterioration.

6.5.3.1 Ethylene Oxide (ETO)

Since many years ago, this chemical has been utilized to reduce the microbial population in spices and herbs. It is remarkably successful at reducing the bacterial population. The operation is complex, though, because there could be health risks for the workers and contamination worries. ETO is allegedly cancer-causing when inhaled but not when consumed with herbs and spices. Regulations in the USA allow for a maximum ETO residue of 50 ppm following treatment. However, European countries forbid using ETO because of potential health concerns. Insects are also killed by ETO treatments at various phases of development, particularly in seedlings of herbs and spices such as celery, cumin, coriander, fenugreek, and fennel that harbour insect eggs as they develop in the field. When these spices are treated with ETO, the eggs are killed and prevented from hatching. After removing the air from the enclosed chamber containing the substance to be treated, pure ETO or an ETO mixture with other gases is delivered. The chamber's remaining ETO is slowly evacuated after a specific time until the level is brought down to the required levels. Herb and spice blends can also be treated with ETO, but they must be devoid of common salt because common salt will react with ETO to produce dangerous chlorohydrins.

Carefully regulating the temperature, duration, and ETO concentration can reduce microbial populations. The substance may only have 50,000 colonies on the plate, 500 yeast and mold colonies, and 10 coliform colonies per gram. Depending on the material type, the residual microbial load will vary. For example, coarse-grained black pepper treated with ETO will have fewer counts than fine-grained black pepper because the ETO gas may permeate the coarsely ground black pepper particles more efficiently than the fine ones. Compared to raw materials with more significant initial counts, those with lower initial counts can attain significantly lower levels following treatment. The raw material's storage container during ETO treatment has an impact on microbial load reduction as well. For instance, the ETO gas penetration and the microbial population suppression are essential if the raw material is packed in burlap bags. There will be less microbial deterioration if similar raw materials are placed in corrugated boxes that can tolerate air evacuation and sealed in thick polythene bags.

6.5.3.2 Propylene Oxide (PPO)

The propylene oxide chemical is liquid with a low boiling point of 34.5°C. Since 1958, it has been used to sterilize food. However, it performs worse than ETO. For the microbiological treatment of spices and herbs, it has been authorized. Due to issues with ETO, many spice companies in California have converted to PPO for paprika and chilli peppers. As with ETO treatment, a vacuum chamber and a volatilizer are necessary for fumigation. PPO has insecticidal qualities as well. A 26-inch-deep mercury vacuum is produced in the chamber where raw ingredients are transformed into vaporized PPO. Four hours later, air cleaning is used to remove the gas. The USFDA sets regulations in CFR 40 Part 185.15, and EPA controls the use of PPO for food fumigation. There must be more than 300 ppm of PPO residue in spices and herbs. Even while the US Environmental Protection Agency (EPA) does not now pose the same worry about handling herbs and spices as ETO, it may ultimately be phased out when ETO is forbidden.

6.5.3.3 Other Chemicals

A particularly effective microbiological strategy for contaminating herbs with chemicals is the use of formaldehyde or ethyl alcohol due to the low efficiency of this method and the low essential oil content of plant materials. This vapour is replaceable by the vapour of ethyl alcohol, which can be checked against using methanol. However, because methanol is harmful, the herbal source material must be completely free of methanol, making this considerably more complex and raising the cost of this method. The advantages and disadvantages of various chemical preservation methods are given in Table 6.4.

6.5.4 OZONE TREATMENT

The best approach for chemically decontaminating plants right now is to utilize ozone. Technology like this is still relatively new. Ozone has solid bactericidal effects sensitive to Gram-negative and Gram-positive bacteria, viruses, yeasts, *Staphylococcus aureus*, and *Listeria monocytogenes*. Bacterial spores are more resistant to ozone than vegetative cells; however, fungi are less sensitive. Because of its potent antioxidant potential, ozone is employed increasingly frequently in the food sector while producing and processing fruits and vegetables. Ozone dissolved in water is used in the recommended herb decontamination techniques; good contact with this component is necessary for the plant's raw material to be effectively purified (Mrozek-Szetela et al., 2020). The storage conditions in gaseous ozone and aqueous ozone are given in Table 6.5.

TABLE 6.4
Advantages and Disadvantages of Various Chemical Preservation Methods

Type of Method	Advantage	Disadvantage
Ethylene Oxide	Kills bacteria, fungi, and viruses very efficiently their spores, too.	Significantly reduces the amount of physiologically active substances, including mucus, glycoside, and alkaloids. Creates carcinogenic chemicals comprise ethylene glycol.
Methyl Bromide	Has a high degree of solubility in organic materials.	Demonstrates a selective impact on the decrease in the raw material containing microbes. Negative impact on the ozone layer on Earth.

TABLE 6.5

Storage Conditions in Gaseous Ozone and Aqueous Ozone

Gaseous Ozone		Aqueous Ozone	
Temperature (°C)	Half-Life (min)	Temperature (°C)	Half-Life (min)
−50	3 months	15	30
−35	18 days	20	20
−25	8 days	25	15
20	3 days	30	12
120	1.5 hours	35	8

6.5.5 Drying Methods

It is reviewed that the degree of binding between mycotoxins and food components, heating temperature, processing time, and heat penetration are the various factors that affect mycotoxins degradation by heat treatment. Typical aflatoxins decompose at temperatures between 237°C and 306°C, which calls for even more serious consideration; solid aflatoxin B$_1$ is stable to dry heating. Once aflatoxins are created, it is not easy to get rid of them with a typical drying process.

Agricultural products must be dried as soon as possible to reduce their moisture content and avoid fungus growth. Until the product reaches safe storage moisture, it should keep drying. In addition, it is critical to avoid using a dryer in a way that promotes microbial growth. It has been suggested to dry in low-humidity conditions to hasten the drying process and enhance the security and quality of the finished product. It must be emphasized once more that aflatoxins cannot be removed using a typical drying procedure once they have formed. Aflatoxin-producing fungi have been prevented from growing by utilizing traditional and cutting-edge drying techniques.

6.5.5.1 Sun or Solar Drying

Following harvest, a standard traditional method is open-air sun drying. Sun drying has comparatively low operating costs compared to other drying techniques but requires lengthy drying durations depending on the temperature. Solar drying is less expensive and more effective than mechanical drying. Solar dryers can produce lower relative humidity and higher drying temperatures than open-air (sun) dryers, which results in lower product moisture content and shorter drying times. A 65% reduction in drying time is observed compared to outdoor drying (Ghazanfari et al., 2003) and developed a thin-layer forced air sun dryer to dry pistachios. A maximum temperature of 56°C, or 20°C over the daily average outdoor temperature, may be reached using the solar collector. Throughout the 36-hour drying process, the product's temperature varied from 15°C to 42°C. Solar drying resulted in items with lighter colours and no aflatoxins as compared to sun drying.

6.5.5.2   Hot Air Drying

The most widely used technique in food and agriculture is hot air drying. This method uses heated air as the medium to provide heat and remove the evaporated moisture. Hot air drying reduces drying time and prevents mold growth compared to solar and sun drying. A more sanitary working environment is also provided by this drying method. According to Yazdanpanah et al. (2005), when pistachios were roasted at 150°C for 30 minutes, the amount of aflatoxin B_1 was reduced by 63%. To speed up the drying process and enhance hygiene, hot air drying of spices at 45°C–600°C was used in place of sun drying.

6.5.5.3   Super-heated Steam Drying

Direct (convective) dryers employ super-heated steam rather than combustion products, hot air, or flue gases to generate drying heat and remove moisture vapour. The technique used for this is super-heated steam drying (SSD).

SSD is superior to hot air drying regarding microbial inactivation and its advantages in generating a dried product with better colour, more nutrients, and porosity. When the material is below the steam saturation temperature at the dryer's operating pressure and comes into contact with super-heated steam, condensation will invariably occur, causing the material's temperature to rise quickly during the early drying phase to the steam saturation temperature. The rapid temperature rise and the material's increased water activity (water content) produce the optimum conditions for the inactivation of microorganisms. SSD could be a fascinating candidate for treating produce to stop mold development and the associated production of aflatoxins, even if no research has been done to assess its potential in this field.

6.5.5.4   Infrared Drying

Infrared radiation (IR) is electromagnetic radiation with wavelengths in the UV and microwave range. The IR spectrum can be divided into the far-infrared (FIR, 3–1,000 μm), middle-infrared (MIR, 1.4–3 μm), and near-infrared (NIR, 0.78–1.4 μm) sections. Foods and IR interact through transmission, absorption, dispersion, and reflection.

NIR produces superior drying outcomes for materials with a thicker body than FIR does for thin layers of a material with a broad surface exposed to radiation. By adding additional sensible warmth to hasten the drying process, IR drying helps to reduce drying time. The substance's composition, moisture content, and structural features impact the penetration depth.

However, aside from aflatoxins, mycotoxins are detoxified by using IR drying. Yılmaz and Tuncel (2010) investigated the effects of drying on grain contaminated with fumonisin. They discovered that the concentration of fumonisins was not considerably affected by either combined IR-hot air drying or IR drying at a temperature of 45°C due to the low drying temperature. The removal of fumonisin requires a higher drying temperature of 60°C.

6.6   CONCLUDING REMARKS

This book chapter highlights the information about dried herbs, spices, and medicinal plants, their storage practices, factors affecting their quality degradation, and precautions to be taken for quality degradation. Due to their low moisture content, they are prone to easy spoilage, eventually leading to quality degradation. Factors affecting quality degradation include internal and external factors such as oxygen, moisture, relative humidity, insects, and birds. While aflatoxins develop, microbial contamination is also the leading cause of degradation in quality. Different storage practices are implemented to improve the shelf life and quality of spices, dried herbs, and medicinal plants. Various packaging materials and technologies such as MAP, CAP, drying, irradiation, and ozone treatment are currently used for advanced storage practices. Physical and chemical methods are also used to prevent microbial growth and insect infestation. In future, utilizing these improved practices will help in better storage, higher shelf life, and improved quality.

REFERENCES

Arora, D. S., & Kaur, J. (1999). Anti-microbial activity of spices. *International Journal of Anti-Microbial Agents*, *12*(3), 257–262.

Ayob, O., Hussain, P. R., Suradkar, P., & Naqash, F. (2021). Gamma irradiation and storage effects on quality and safety of Himalayan paprika (Waer). *LWT*, *147*, 111667.

Barnes, J., Anderson, L. A., & Phillipson, J. D. (2007). *Herbal Medicines*. Pharmaceutical Press.

Bhat, R., Rai, R. V., & Karim, A. A. (2010). Mycotoxins in food and feed: present status and future concerns. *Comprehensive Reviews in Food Science and Food Safety*, *9*(1), 57–81.

Chatterjee, S., Padwal-Desai, S., & Thomas, P. (1998). Effect of γ-irradiation on the colour power of turmeric (Curcuma longa) and red chillies (Capsicum annum) during storage. *Food Research International*, *31*(9), 625–628.

Costa, L. C. B., Pinto, J. E. B. P., Bertolucci, S. K. V., Alves, P. B., & Evangelino, T. S. (2009). Variation in essential oil yield and chemical composition of whole and powdered "atroveran" (Ocimum selloi Benth.) leaves under storage conditions. *Revista Brasileira de Plantas Medicinais*, *11*, 43–48.

Deshwal, G. K., Panjagari, N. R., & Alam, T. (2019). An overview of paper and paper based food packaging materials: health safety and environmental concerns. *Journal of Food Science and Technology*, *56*(10), 4391–4403.

George, C. (2006). Other decontamination techniques for herbs and spices. In K. V. Peter (ed.), *Handbook of Herbs and Spices* (pp. 74–85). Elsevier.

Ghazanfari, A., Tabil Jr, L., & Sokhansanj, S. (2003). Evaluating a solar dryer for in-shell drying of split pistachio nuts. *Drying Technology*, *21*(7), 1357–1368.

Iqbal, Q., Amjad, M., Asi, M. R., & Arino, A. (2012). Mold and aflatoxin reduction by gamma radiation of packed hot peppers and their evolution during storage. *Journal of Food Protection*, *75*(8), 1528–1531.

Jung, K., Song, B.-S., Kim, M. J., Moon, B.-G., Go, S.-M., Kim, J.-K., Lee, Y.-J., & Park, J.-H. (2015). Effect of X-ray, gamma ray, and electron beam irradiation on the hygienic and physicochemical qualities of red pepper powder. *LWT-Food Science and Technology*, *63*(2), 846–851.

Kamat, A., Pingulkar, K., Bhushan, B., Gholap, A., & Thomas, P. (2003). Potential application of low dose gamma irradiation to improve the microbiological safety of fresh coriander leaves. *Food Control*, *14*(8), 529–537.

Kim, J. H., Shin, M.-H., Hwang, Y.-J., Srinivasan, P., Kim, J. K., Park, H. J., Byun, M. W., & Lee, J. W. (2009). Role of gamma irradiation on the natural antioxidants in cumin seeds. *Radiation Physics and Chemistry*, *78*(2), 153–157.

King, K. (2006). Packaging and storage of herbs and spices. In Peter, K. V. (ed.), *Handbook of Herbs and Spices* (pp. 86–102). Elsevier.

Mayuoni‑Kirshinbaum, L., Daus, A., & Porat, R. (2013). Changes in sensory quality and aroma volatile composition during prolonged storage of 'Wonderful' pomegranate fruit. *International Journal of Food Science & Technology, 48*(8), 1569–1578.

Mrozek-Szetela, A., Rejda, P., & Wińska, K. (2020). A review of hygienization methods of herbal raw materials. *Applied Sciences, 10*(22), 8268.

Munasiri, M., Parte, M., Ghanekar, A., Sharma, A., Padwal‑Desai, S., & Nadkarni, G. (1987). Sterilization of ground prepacked Indian spices by gamma irradiation. *Journal of Food Science, 52*(3), 823–824.

Nafiu, M., Hamid, A., Muritala, H., & Adeyemi, S. (2017). Preparation, standardization, and quality control of medicinal plants in Africa. In Victor Kuete (ed.), *Medicinal Spices and Vegetables from Africa* (pp. 171–204). Academic Press.

Nura, A. (2018). Advances in food packaging technology-a review. *Journal of Post-Harvest Technology, 6*(4), 55–64.

Plusquellec, A. (1995). In Bourgeois, C. M., & Leveau, J. Y. (eds.), *Microbiological Control for Foods and Agricultural Products* (pp. 437–443). John Wiley & Sons Inc.

Rahman, M., Islam, M., Das, K. C., Salimullah, M., Mollah, M., & Khan, R. A. (2021). Effect of gamma radiation on microbial load, physico-chemical and sensory characteristics of common spices for storage. *Journal of Food Science and Technology, 58*(9), 3579–3588.

Singh, A., Sharma, P. K., & Malviya, R. (2011). Eco friendly pharmaceutical packaging material. *World Applied Sciences Journal, 14*(11), 1703–1716.

Šojić, B., Tomović, V., Kocić-Tanackov, S., Škaljac, S., Ikonić, P., Džinić, N., Živković, N., Jokanović, M., Tasić, T., & Kravić, S. (2015). Effect of Nutmeg (Myristica fragrans) essential oil on the oxidative and microbial stability of cooked sausage during refrigerated storage. *Food Control, 54*, 282–286.

Song, W.-J., Sung, H.-J., Kim, S.-Y., Kim, K.-P., Ryu, S., & Kang, D.-H. (2014). Inactivation of Escherichia coli O157: H7 and Salmonella Typhimurium in black pepper and red pepper by gamma irradiation. *International Journal of Food Microbiology, 172*, 125–129.

Steinhart, C. E., Doyle, M. E., & Cochrane, B. A. (1995). *Food Safety 1995*. CRC Press.

Topuz, A., & Ozdemir, F. (2004). Influences of gamma irradiation and storage on the capsaicinoids of sun-dried and dehydrated paprika. *Food Chemistry, 86*(4), 509–515.

Verma, M., Shakya, S., Kumar, P., Madhavi, J., Murugaiyan, J., & Rao, M. (2021). Trends in packaging material for food products: historical background, current scenario, and future prospects. *Journal of Food Science and Technology, 58*(11), 4069–4082.

World Health Organization. (1999). *The World Health Report: 1999: Making a Difference*. World Health Organization.

Yazdanpanah, H., Mohammadi, T., Abouhossain, G., & Cheraghali, A. M. (2005). Effect of roasting on degradation of aflatoxins in contaminated pistachio nuts. *Food and Chemical Toxicology, 43*(7), 1135–1139.

Yılmaz, N., & Tuncel, N. B. (2010). An alternative strategy for corn drying (Zea mays) resulted in both energy savings and reduction of fumonisins B1 and B2 contamination. *International Journal of Food Science & Technology, 45*(3), 621–628.

7 Product Development and Formulations from Dried Herbs, Spices, and Medicinal Plants

Fatemeh Poureshmanan Talemy,
Fereshteh Jamalzade, Kosar Mohammadi Balili,
and Narjes Malekjani
University of Guilan

CONTENTS

DOI: 10.1201/9781003269250-7

7.1 INTRODUCTION

The application of medicinal plants and herbs has been raised in the modern world to cure numerous illnesses with much less poisonous effects. Nowadays, herbal products are available in various forms in the market, including liquid preparation such as teas, tinctures, etc., as well as powders such as dried extracts in tablets or capsules. Novel drugs are now a prominent part of the modern world. Special attention has been given to developing herbal and medicinal plant formulations using novel methods and technologies due to the exceptional demand for these products in evolved countries.

Several methods are used to develop herbal formulations, divided into traditional and novel techniques. The conventional methods used to prepare herbal drugs and medicines include simple procedures such as hot- or cold water extraction, exudation of juice, and drying the formulation powder into pastes with water, oil, or honey, and even fermentation after the addition of a sugar source. Traditional methods have several disadvantages; thus, novel technologies have been used to modify these procedures and improve the drug's effectiveness and stability with a better pharmacokinetic route for targeted delivery.

Improved bioavailability, reduced toxicity with higher stability, solubility, better tissue macrophage distribution, and sustained delivery are the notable benefits of novel formulations over traditional ones (Sarangi & Padhi, 2018). The novel methods used for developing herbal and medicinal plant formulations include the application of drug delivery systems with controlled and targeted release mechanisms. Drug delivery systems with controlled mechanisms represent numerous advantages compared to traditional forms. Artificial polymers are generally utilized as carriers in such structures to deliver the drug at a specific time, which is extremely useful for medicines with higher drug metabolism. Several mechanisms have been developed for controlled release applications from temporary and dispensation delivery systems. Different drugs impose distinct limitations according to the type of delivery system used, so the variation of these methods is essential. For instance, the medication that is going to diffuse into the stomach, an environment that contains high acidity and considerable instability, will undoubtedly possess a different delivery system from that of an intravenous delivery (Uhrich, Cannizzaro, Langer, & Shakesheff, 1999). Targeted release systems operate by directing the drug into the aimed tissue leading to its absorption and preventing it from being released along the way.

The herbal medicine delivery systems based on materials with the size of nano are a new generation of formulation with improved activity and overcoming problems related to plant medicines (Saraf, 2010). The quality control of the prepared herbal drugs and their standardization could be mentioned as another significant step in their preparation procedures, which includes various parameters and examinations that will be indicated in this chapter.

Regarding all aforementioned contents, the present chapter aimed to address an overview of various dried herbs and spices along with their preparation methods as well as ways for developing herbal and medicinal plants formulations, packaging, labelling, storage, and transportation conditions are also discussed in the following contents.

7.2 PRODUCTION OF DRY HERBS AND SPICES

7.2.1 Cloves

The clove *Syzygium aromaticum (L.) Merrill et.* Perry pertains to the family *Myrtaceae*. The cloves contain volatile oils such as eugenol (up to 95%), acetyl eugenol, B-caryophyllene, vanillin, etc. The plant is harvested for the closed flower buds.

  Dried cloves production: After harvesting, the cloves are dried. The drying is done by the sun or an artificial dryer.

7.2.1.1 Ground Cloves Production

Ground clove is produced by grinding dry clove buds. The temperature is constrained to 25°C–35°C for preserving volatile compounds. The techniques include cryogenic grinding or water-cooled grinders.

7.2.1.2 Main Uses of Cloves

The clove is widely used for food preservatives, spices, herbal tea, and pharmaceuticals. Clove herbal tea is used to treat nausea and help digestion and relieve pain.

7.2.2 Basil

Basil (*Ocimum basilicum*, Lamiaceae) is a spicy herb native to India.

7.2.2.1 Dried Basil Production

The leaves are harvested before flowering while they are fresh. Thereafter, leaves are washed and cleaned to eliminate extraneous materials, followed by air-drying the leaves in a warm room or placed on a drying rack. The leaves are removed from the stems and kept in a dark and cool place. It is preferred to keep the dry leaves below 30°C under reduced pressure conditions (K. V. Peter & Babu, 2012).

7.2.2.2 Main Uses of Basil

The main uses include flavouring purposes in various food and herbal tea. The essential oil of *O. basilicum* gained by distillation is utilized as a flavouring agent correlated to its aromatic characteristic. Also, it can be used in non-alcoholic beverages. Sweet basil is used to treat headaches, cough, inflammation, snakebite, and rabies (P Pushpangadan, Rajasekharan, & Biju, 1993).

7.2.3 Mint

Mints are aromatic herbs that belong to the Lamiaceae family and genus *Mentha*. The essential oil of mint, a rich source of menthol, is used in food and medicinal industries.

7.2.3.1   Dried Mint Production

Harvesting is an important step during production. Harvesting at an unsuitable time leads to producing essential oil of low quality. Japanese mint should be harvested 100–110 days after planting. Peppermint and spearmint should be harvested 90–95 days after cultivation. After harvesting, the undesired stems should be removed, and the herb must be washed. Thereafter, the excess water is removed by bloating paper. The peppermint is then dried by laying in single (1 cm), double (2 cm), and three layers (3 cm) inside the solar tunnel dryer. The solar tunnel dryer performance could be enhanced by using a solar photovoltaic system and a smooth plate collector for drying (Eltawil, Azam, & Alghannam, 2018). The drying is continued until the weight loss is fixed. After drying, the dried herbs are stored.

7.2.4   CINNAMON

Cinnamon is an ancient spice. The genus *Cinnamomum* belongs to the family Lauraceae. The commercial product of cinnamon is obtained from the inner bark of the tree *Cinnamomum Verum*. Two or 3 years after planting, the barks are harvested (P. Pushpangadan & George, 2012).

7.2.4.1   Production of Quills

Peeling: By using a small machine (RUWEKA-CG rubbing device), the barks are removed from cinnamon stems.

Rolling: The barks are stacked together and pressed well and coated with a mat in order to moisture retention to facilitate piping and also to increase the slight fermentation.

Piping: The outer layer is scratched by using a curved knife. The end of the slips is cut and pressed. This process generates pipes (P. Pushpangadan & George, 2012).

Drying: The pipes are dried under the shade. Direct exposure to the sun would lead to the wrapping of pipes. After drying, the colour of the quills turns to yellowish-brown (P. Pushpangadan & George, 2012).

7.2.4.2   Production of Ground Cinnamon

Ground cinnamon is commonly used as a spice. During grinding, an amount of heat is generated which can lead to decrease volatile oil content of cinnamon. Thus, cryogenic grinding is a suitable method to preserve more volatile oils (Tainter & Grenis, 1993).

7.2.5   ROSEMARY

Rosemary (*Rosmarinus officinalis* L.) is an ancient herb to the Lamiaceae family. It is indigenous to the Mediterranean region.

7.2.5.1   Dried Rosemary Production

Harvesting occurs until 50% of flowering. After harvesting, the dehydration is carried out using dry air at 30°C temperatures. This method can preserve the quality and antioxidant properties of rosemary. The dried rosemary can be used in the form of herbal tea, spices, medicinal, and cosmetics (P. Pushpangadan & George, 2012).

7.2.6 THYME

The genus *Thymus* belongs to the Labiate family tribe Mentheae. This herb is native to the Mediterranean region.

7.2.6.1 Dried Thyme Production

The harvest occurs during the blooming time. The period of blooming can be different in various regions. The leaves and flowers are harvested. Harvesting time is an important factor that affects the essential oil content. One of the techniques of drying is the utilization of hot dry air. Venskutonis (2002) employed cross-flow and through-flow drying methods on Indian thyme at 40, 50, and 60°C and reported that drying at 40°C provided the best consequences. Another drying method appertains to be freeze-drying. This method can preserve the colour, aroma, and quality. The thyme is used as a spice, herbal tea, antioxidant agent, and for pharmaceutical purposes.

7.2.7 SAFFRON

Saffron (*Crocus sativus* Linnaeus) pertains to the Iridaceae family and genus *Crocus*. It is the most valuable spice. Saffron is found mostly in Mediterranean Europe and Western Asia.

7.2.7.1 Dried Saffron Production

To produce the saffron, the stigmas are harvested. Depending on the countries, there are various methods for drying. For instance, in Italy, the stigmas are placed on oak embers, which take 15–20 minutes to dry. The drying continues until the moisture content is between 5% and 20% (P. Pushpangadan & George, 2012). In Iran and India, drying mostly occurs through traditional sun drying. In India, stigmas are dried with direct exposure to the sun for 3–5 days till the moisture content reaches 8%–10% (Negbi, Dagan, Dror, & Basker, 1989; Sampathu, Shivashankar, Lewis, & Wood, 1984). Saffron is a beneficial herb that is used as a spice and herbal tea. Also, it is utilized for medicinal purposes such as anti-inflammatory, anti-spasmodic, and sedative.

7.2.8 TURMERIC

Turmeric of commerce is the dried rhizome of the plant *Curcuma domestica*. It belongs to the family *Zingiberaceae*. Turmeric is grown in Asian countries, especially India being the leading producer of turmeric in the world.

7.2.8.1 Dried Turmeric Production

After harvesting, turmeric is washed to remove extraneous materials. Mothers and fingers are separated and boiled for about 1 hour under slightly alkaline conditions. Afterwards, the turmeric is solar-dried or tunnel dried until the moisture content

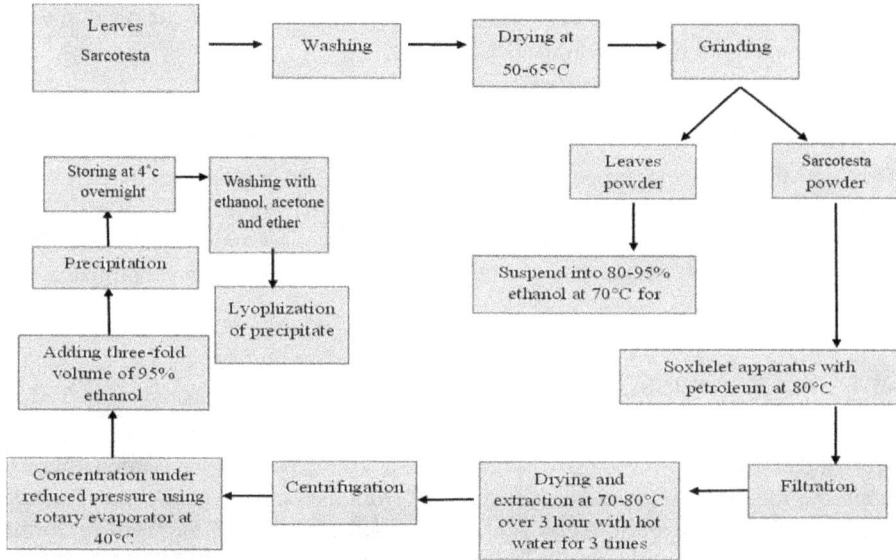

FIGURE 7.1 Extraction and purification of *G. biloba*.

is decreased to 10%–20%. After drying, the dried turmeric is polished to improve colour and marketability. The turmeric milling is performed by hammer mill to obtain 60–80 mesh powder. Turmeric is a commonly used spice and also can be used for medicinal purposes such as treatment for vertigo, sprains, diarrhoea, and also it is effective in healing wounds (P. Pushpangadan & George, 2012).

7.2.9   *G. BILOBA* LEAVES

Ginkgo biloba L., also known as *yinhsing* (Chinese), is a perdurable plant in the Ginkgoaceae family. It is a traditional Chinese herb used for around 250 million years for medical purposes in China, Japan, and Korea (Hatano, Miyakawa, Sawano, & Tanokura, 2011; Singh, Kaur, Singh, & Ahuja, 2008). Figure 7.1 demonstrates the extraction and purification of *G. biloba* (Fang, Wang, Wang, & Wang, 2020).

7.2.10   GINSENG

Ginseng Radix et Rhizome (Renshen, in Chinese) is one of the most well-known herbs in Asian countries. Triterpene saponin is the main active agent in ginseng that causes anti-diabetic, immune-modulatory, anti-cancer, anti-inflammatory, and anti-oxidant properties (Siraj, Sathish Kumar, Kim, Kim, & Yang, 2015).

7.2.10.1   Dried Ginseng Production

The two types of ginseng include white ginseng (fresh air-dried) and processed one, Hongshen, which is steamed for 2–3 hours and dried. Recent studies have shown that Hongshen is more efficient than white ginseng (Siraj et al., 2015). During steaming, the malonyl ginsenosides are converted to ginsenosides, and this chemical change generates beneficial properties (Siraj et al., 2015).

7.3 FURTHER PROCESSING OF HERBS AND SPICES

Herbs can also be further processed to produce products other than dried herbs. Such products are more ready to use. Extraction of the herbal extract and converting the extract to the form of powders or granules is one of the most widely used methods to produce ready-to-use herbal teas and other products which will be explained in the following sections.

7.3.1 SIZE REDUCTION

After the initial drying of herbs (Chapters 2 and 3), the size reduction is performed before the extraction process. Size reduction would increase the surface area, thereby increasing the extraction yield of active agents from the herbs. Active agents include flavonoids, tannin, saponin, gallic acid, polyphenols, etc. A hammer mill performs milling to reach the optimal size between 30 and 40 mesh (Selvamuthukumaran, Tranchant, & Shi, 2019). Spice grinding is accomplished to gain higher taste and flavour characteristics. The particle size of the ground spice powders could be different from 50 to 850 µm. The particle size is subjugated to its end-use. A hammer mill, attrition mill, or pin mill is utilized for spice grinding. One of the best grinding methods is cryogenic grinding technology, which uses a very cold gas in the form of liquid at a temperature below −153°C. Cryogens comprise liquid nitrogen, helium, carbon dioxide, argon, etc. The use of cryogenic grinding for spices increases their flavour and therapeutic properties (Saxena et al., 2018). In the conventional grinding of spices, some heat is generated. This heat could reduce the quality and quantity of the product. Depending on the oil and moisture content in the seeds, the volatile oil content might be reduced up to 30%–40%. Besides, the colour of products might decrease. Therefore, cryogenic grinding of spices is a preferred choice to maintain the colour of the product during grinding. The volatile oil is considerably more (2.61%) in cryo-ground black pepper in contrast to 1.5% in conventional grinding technology (Jacob, Kasthurirengan, Karunanithi, & Behera, 2000). Air pollution occurs during conventional grinding, which is addressed by cryogenic grinding. Also, in conventional grinding, microbial contamination is comparatively higher than cryogenic grinding.

7.3.2 EXTRACTION

The selection of the solvent for the extraction depends on the nature of active agents. Purification and recognition of the bioactive compound are required to select an optimal extraction technique. Due to the complex structure of plants and the different physiochemical properties of bioactive compounds, extraction could be difficult. Conventional extraction (CE) includes hydrodistillation with cohobation, maceration, enfleurage, percolation, infusion, and decoction (Selvamuthukumaran, 2019). CE is an inexpensive extraction method that intends to extend the solubility of desired compounds while increasing the mass transfer. The non-conventional extraction includes ultrasound-assisted extraction, enzyme-assisted extraction, microwave-assisted extraction, super-critical fluid extraction, high-pressure-assisted extraction, and electrically assisted extraction (Giacometti et al., 2018; Selvamuthukumaran, 2019).

7.3.2.1 Conventional Extraction

7.3.2.1.1 Hydrodistillation with Cohobation

In the industrial process of hydrodistillation, the stills are filled with herbs and water and then steam-heated for about 90 minutes. The evaporated water and herb essential oils are removed from the still, condensed, and collected. The oil separated is called direct oil, which includes 20% of the total oil. The water which condenses along with the oil is unloaded and distilled again (cohobation) to achieve the water-soluble fractions. After cohobation, the achieved oil is called indirect oil which contains 80% of the total oil. The direct and indirect oil mixture creates commerce herbal oils (Erbas & Baydar, 2016).

7.3.2.1.2 Maceration

In this method, the herbs are steeped in the solvent in a closed container. The container is stored at room temperature for at least 3 days with repeated agitation to rupture the plant cell wall to release the soluble material. After 3 days, the mixture is passed through the filter. The solvent evaporates through heating. After evaporation, the extract is dried by spray drier to a powder form. The advantages of this method are low cost and ease of operation. The disadvantages of this method include longer process time, higher solvent requirement, and degradation of active agents due to high temperature (Selvamuthukumaran, 2019).

7.3.2.1.3 Percolation

In this technique, the plant material is placed in a thimble holder. Using a pump, the solvent and extract combination is re-spread in the system. The advantage is the possibility of selecting an operating temperature and no need for filtration. The main disadvantage is that the returned solvent is not fresh, reducing solute recovery (Leal, Almeida, Prado, Prado, & Meireles, 2011).

7.3.2.1.4 Infusion and Decoction

This technique is similar to maceration. In both of them, the herbs are soaked in a solvent, but the advantage of this method is that it takes less time compared to maceration. The size reduction is performed primarily to extract heat-resistant constituents and complex components such as roots and barks (Selvamuthukumaran, 2019).

7.3.2.1.5 Enfleurage

In this method, dry herbs are pounded with mortar and pestle to improve the absorption area. After pounding, the mashed herbs are mixed with warmed oil to absorb the essential oil. Afterwards, the container is covered with aluminium foil and allowed to stand for 24 hours at room temperature. Then, ethanol is added to absorb the essential oil. Ethanol is then separated from the essential oil (Parab, Salgaonkar, Padwekar, & Purohit).

7.3.2.2 Non-conventional Extraction

7.3.2.2.1 Ultrasound-Assisted Extraction

In this method, the waves of a frequency ranging between 20 kHz and 2,000 MHz are used. Ultrasonication produces cavitation once the acoustic power input is sufficiently high to permit multiple micro-bubbles at nucleation sites within the fluid. The bubbles grow throughout the rarefying phase of the acoustic wave and then collapse during the compression phase (Chemat & Strube, 2015). The ultrasound waves lead to cavitation that can facilitate the permeation of solvent into the plant samples (Selvamuthukumaran, 2019). This technique leads to cell wall breakdown and enhances the desorption of bioactive components from the solid matrix (Gouda et al., 2021). The disadvantage of this method is the production of free radicals from energetic ultrasound waves, leading to bioactive degradation.

7.3.2.2.2 Enzyme-Assisted Extraction

Using enzymes such as amylases, cellulases, and proteases breaks down the cell walls of plant material and can decrease particle size. Therefore, the active components are released from the sample. Enzymes can be adsorbed onto a substrate for recovery after extraction (Chemat & Strube, 2015).

7.3.2.2.3 Microwave-Assisted Extraction

This technique offers rapid extraction from plant materials. The benefits include lower solvent consumption, lower time consumption, and reduced environmental pollution. Electromagnetic waves with a frequency between 300 MHz and 300 GHz are named microwaves. In the microwave process, the couple ionic conduction and dipole rotation lead to the sample heating up. Thus, warming the sample and creating significant pressure on the cell wall. Consequently, the cell wall weakens and becomes susceptible to breakdown. It helps in enhancing the extraction yield of phytoconstituents. The extraction yield could be improved by suitable solvent selection and elevated temperature, accelerating solvent penetration in the cell wall.

7.3.2.2.4 Super-critical Fluid Extraction

Super-critical extraction is a method commonly used with carbon dioxide. The CO_2 is a suitable solvent because of its lesser toxicity, inexpensive, easy to use, and generally recognized as safe (GRAS) (Ruttarattanamongkol, Siebenhandl-Ehn, Schreiner, & Petrasch, 2014). In this technique, the sample is contained in the extraction vessel, and CO_2 is transferred into the vessel. CO_2 dissolves the extract from the sample matrix. Then, the sample is transferred into the collector, and the extract is collected by releasing the pressure (Selvamuthukumaran, 2019).

7.3.2.2.5 High-Pressure-Assisted Extraction

This method consists of three stages: the come-up, pressure maintenance, and pressure relief. The applied pressure in the extraction process is about 100–1,000 MPa. The sample is placed in a pressurized vessel. Due to the pressure applied in the maintenance stage, the force of both sides of the cell wall is balanced and allows the solvent to penetrate through the cell wall and dissolve the active components

(Giacometti et al., 2018). After the extraction process, the mixture is percolated and dried by utilizing a rotary evaporator.

7.3.2.2.6 Pulsed Electric Field (PEF) Extraction

PEF strength ranges from 15 to 35 kV/cm and energies of 50–700 kJ/kg. In these circumstances, the cellular membrane of plants is electroporated (T. Wang et al., 2017). Recent studies have recognized PEF to be appropriate for extraction owing to its ability to free intra-cellular species with lower energy consumption (Parniakov, Barba, Grimi, Lebovka, & Vorobiev, 2016). The advantages of PEF are shortened extraction time, better extraction, and lower temperature consumption (Puertolas, Koubaa, & Barba, 2016; Yan, He, & Xi, 2017). The other novel technology is high-voltage electric discharge plasma (HVED) which is known as corona discharge. This technology, with pulsed quick discharge voltages from 20 to 80 kV/cm, enables the breakdown of the cell wall and membrane of plant material (Boussetta & Vorobiev, 2014).

7.3.3 SOLVENT RECOVERY

The solvent is separated by a heating component and a condensation component to condense the liquid and separate the solvent from miscella. This technique may also require vacuum equipment that can evaporate the solvent at a lower temperature (Selvamuthukumaran, 2019).

7.3.4 DRYING OF HERBAL EXTRACTS

Herbal extract drying techniques encompass spray drying, freeze-drying, and spouted bed drying. But drying herbal extracts using a spray drier is a more common method in the herbal processing industry (Souza & Oliveira, 2004).

In this process, the extract containing 5%–10% dry material enters the spray tower. Dry air of 120°C–350°C is fed. The atomization occurs at a pressure ranging from 10 to 150 bar by pressure nozzles (Berkulin & Theissing, 2003). After that, the extract is converted to powder form.

7.4 DRUG DELIVERY SYSTEMS

Novel delivery systems for herbal drugs and medicines have gained significant attention in recent decades. The application of herbal medications to cure numerous illnesses with less toxic properties and better therapeutic efficacy has become outstanding in the present day. Meanwhile, some limitations, such as instability in highly acidic pH, are faced when using herbal extracts/plant actives (Sarangi & Padhi, 2018). The degradation of herbal drugs and their side effects due to the accumulation of herbal medicines in undesired areas might be minimized by the modulation of novel delivery technologies.

Some critical needs of novel carriers, for example, the power to deliver the medication at an appropriate rate or to transfer these medicines to the desired tissue, might not be satisfied by the conventional dosages. Natural products must consist of both hydrophile and hydrophobic parts in their chemical structure for good bioavailability and higher permeability through lipidic membranes with better solubility in gastrointestinal liquids. Several herbal compounds, such as polyphenols, have better dissolve in water wherein lack lipophilicity. There could be two reasons; either because of their multiple-ring and large particles or their limitation in reaching the lipophilic outer membranes due to their negligible miscibility with different lipids. By targeting the affected area particularly and transporting the drug to that section, novel carriers could treat specific diseases (Sarangi & Padhi, 2018).

Several approaches have been used in novel drug delivery systems and their carriers. Different types of novel herbal drug delivery systems are introduced in the following sections.

7.4.1 LIPOSOMES

Liposomes, which are a subset of surfactant-based delivery systems, are one of the structures studied extensively. These lipophilic carriers are generally used in food and drugs. They can provide various brilliant features such as enhancing the stability and tolerance of the drug, biological affiliation, the ability to target a specific cell or zone, and low toxicity. The liposomes can encapsulate a fraction of the solvent into their spherical concentric bilayer structures, in which the non-polar tails of the molecules face each other. Due to the constituent number of layers of liposomes, uni-lamellar and multi-lamellar liposomes are available, which slightly differ in physical structure. As shown in Figure 7.2, liposomes can encapsulate different kinds of active compounds in one delivery system due to the hydrophilic and hydrophobic parts in their molecular structure (McClements, 2014). Hydrophobic drugs can be trapped between the surfactant tails in the bilayer region, while hydrophilic compounds can be trapped in the innermost aqueous section.

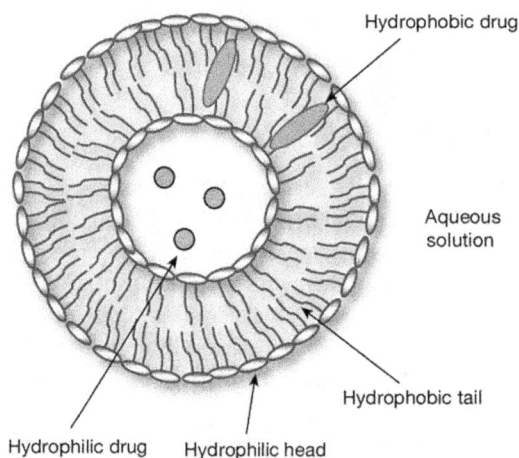

FIGURE 7.2 Encapsulation of drugs by liposomal delivery systems. (From Lee (2020) with permissions.)

7.4.2 PHYTOSOMES

Phytosomes or phytolipid delivery systems are lipid-compatible molecular complexes, and they are used for poorly bioavailable and water-soluble constituents of phyto-medicines to enhance their absorption and bioavailability. They can also improve stability profiles by forming a stable complex with phospholipids. Phytosomes can pass better through the lipophilic surroundings, entering the cell membrane and eventually reaching the blood/targeted origin, and in the meantime, they protect the beneficial components of herbal extract from stomach bacteria and digestive secre-tions. Phospholipids from soy, mainly phosphatidylcholine, are used as the lipid-phase substance for phytosome composition. For example, phytosomal formulation of curcumin, which is an anti-cancer and antioxidant plant constituent, is used to increase its bioavailability and antioxidant activity (Jadhav, Wadhave, Arsul, & Sawarkar, 2014).

7.4.3 NANOPARTICLES

Nanoparticles (Figure 7.3) are efficacious delivery systems for hydrophobic and hydrophilic drugs. The size of nanoparticles commonly ranges between 10 and 1,000 nm (Goyal et al., 2011). In recent years, considerable attention has been given to biodegradable polymeric nanoparticles as a conceivable drug delivery system. Increasing the solubility of the compounds and enhancing the absorption of drugs while reducing their dosages compared to their conventional counterparts are some of the numerous benefits of using nanotechnology as a delivery system. The nano-spheres structure contains a matrix in which the active ingredients of the drugs are dispersed. In comparison, a polymeric membrane surrounds the active ingredient (as a core) in the nanocapsules. An example is curcuminoid solid-lipid nanoparticles, which prolong the release of curcuminoids as an antioxidant and anti-cancer com-pound (Mukerjee & Vishwanatha, 2009).

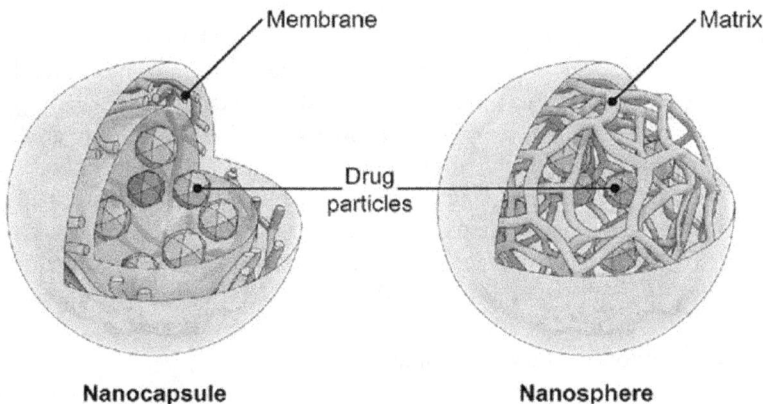

FIGURE 7.3 Encapsulation of drugs by nanocapsule and nanosphere delivery systems. (From Rambaran (2020) with permissions.)

7.4.4 NIOSOMES

Niosomes are created from non-ionic surfactants (of the alkyl or dialkyl polyglyc-erol ether class) along with cholesterol and are either uni-lamellar or multi-lamellar (Sarangi & Padhi, 2018). Their bilayer design is constructed by the selfassembly of hydrated surfactant monomers. The earlier studies demonstrated that niosomes have similar possessions to liposomes, but because of their non-ionic surfactant, niosomes are more stable than liposomes. Thus, they can overcome liposome-related problems, i.e., susceptibility to oxidation, high price, and difficulty procuring high purity lev-els, which influence stability, shape, and size. Niosomes have bilayer structures with hydrophilic outer and inner surfaces, while the area between the layers is of hydropho-bic compounds for the delivery of active compounds and medications (Sankhyan & Pawar, 2012).

7.4.5 PRONIOSOMES

Proniosomes are gel systems used for several applications in drug delivery systems that target a specific area. Before using proniosomes in hot aqueous media, they have to rehydrate to form niosomal dispersions instantly. Proniosomes are physically stable during transport and storage, with enhanced drug penetration at the target tis-sue, thereby reducing toxicity (Akhilesh, Faishal, & Kamath, 2012).

7.4.6 BIOPOLYMERS USED IN HERBAL DELIVERY SYSTEMS

There are many examples of biopolymers, which are generally assembled from pro-teins and polysaccharides, used to devise food-grade colloidal dispersions appro-priate for encapsulation and delivery of active components such as herbal extracts including casein, gelatin, soy protein, zein, cellulose, starch, whey proteins, and many other hydrocolloids. Molecular characteristics of polymers, including flexibil-ity, conformation, lipophilicity, molecular weight, reactivity, and molecular charge, typically concern the ability of a biopolymer to form colloidal particles for improved functional performance and physicochemical properties. These molecular properties are impacted by numerous variables, including the nature of intra-molecular bonding and their type and distribution along the chain (McClements, 2014).

Biopolymers used for commercial applications must be designed to contain spe-cial functions and physicochemical features for specific applications. For example, a biopolymer used for delivering a flavour ingredient is better to be formed in such a way as to release the encapsulated molecules in the mouth. In return, a biopoly-mer meant to deliver an anti-cancer ingredient to the colon must remain undisrupted while passing the mouth, stomach, and small intestine but breakdown and release when reaching the colon. The following is a brief overview of some of the food-grade biopolymers commonly used to assemble delivery systems.

7.4.6.1 Milk Proteins

Milk proteins are segregated into two major classes known as caseins and whey pro-teins. Both groups consist of a complex mix of different proteins, where the possibility of fractionating them into discrete purified protein fractions such as β-lactoglobulin, α-lactalbumin, etc., is established. Generally, caseins in milk or 'casein micelles' consist of a group of submicelles held together by mineral ions such as calcium phosphate. They usually vary between 50 and 250 nm in diameter. These micelles can solubilize hydrophobic ingredients such as carotenoids, oil-soluble vitamins, and curcumin in their internal structure, equipping themselves as natural biopolymer-based delivery systems.

Native whey protein molecules include a blend of primarily globular proteins that vary in molecular specifications. In a wide range of pHs, they are highly soluble in water, although they may aggregate near their isoelectric point. Especially, the oil droplets coated by whey proteins are so sensitive due to the reduction of their electrostatic excretion near the isoelectric point. While heating the whey proteins above a critical temperature, their structure would unfold, which could be diverse depending on the fractions. When they unfold, the amino acid side groups located in the inner part become exposed, and their reactive nature causes the whey proteins to aggregate by cross-linking. This structural change may be detrimental or beneficial depending on the purpose of its application (McClements, 2014).

7.4.6.2   Meat and Fish Proteins

Many proteins in meat and fish can be utilized to assemble drug delivery systems based on biopolymers, for instance, gelatin, which is used extensively (Phillips & Williams, 2011). Gelatin is a protein of animal origin (pig, cow, or fish) with a relatively high molecular weight and is isolated from collagen. Commercially, gelatin is prepared by boiling the collagen with a strong acid or alkaline so its structure will be denatured, which results in the production of type A and type B gelatin according to the chemical agent used during the boiling process. Due to their different preparation methods, the isoelectric point of type B gelatin is lower than type A. Gelatin is mainly used in foods for thickening and gelation but can also be used as a coating for oil droplets in emulsions as a result of having an active surface (McClements, 2014).

7.4.6.3   Egg Proteins

Biopolymer-based delivery systems can be fabricated with the globular proteins of egg yolk and egg white. The application of these proteins is similar to the previously mentioned proteins. Hence, they can stabilize oil-in-water (O/W) emulsions due to their active surface. Also, heating these proteins over their denaturation temperature can form gels and cause solutions to become thicker. Eventually, they can be utilized as structuring agents to develop novel designs in delivery systems, considering their electrostatic and hydrophobic interactions with other molecules.

7.4.6.4   Plant-Based Proteins

Various plant species contain proteins used to create delivery systems, for example, cereals and legumes. Soy protein is one of the extensively used plant proteins and includes a complex blend of globular proteins. Soy proteins may be used as structure formers, thickening, gelling agents, and emulsifiers, as they are surface-active like the rest of the proteins mentioned above (McClements, 2014).

Zein is another protein used a lot and is derived from corn. It is also very hydrophobic. This protein can be used to fabricate colloidal particles to encapsulate active components, which are lipophilic, due to their ability to be soluble in concentrated ethanol solutions. At the same time, their solubility in pure water is limited (McClements, 2014).

7.4.6.5   Starch

Various sources such as wheat, corn, tapioca, potato, and rice can be used to derive starch. Starch consists of two main molecular structures named amylose and amylopectin. Amylose is a linear molecule which is formed of glucose units linked by α-D-(1–4) bond. However, amylopectin has a highly branched structure composed of a primary linear molecule that consists of α-D-(1–4) linked glucose units linked with some limited branches by α-D-(1–6) bond. The ratio of amylopectin and amylose in the starch alters according to its diverse sources (McClements, 2014).

The starch granules available in nature include two different regions, a crystalline region which is separated by an amorphous region (Pérez & Bertoft, 2010), which becomes evident on heating the starch above a critical temperature. It absorbs water causing the crystalline area to disrupt and causing the starch to swell, leading to a considerable increase in the viscosity of the solution. If the concentration of starch in the solution is high enough, it may lead to gelation, called 'gelatinization'. After the starch solution is cooled, linear structures link together through hydrogen bonds, which results in the thickening of the solution, known as 'retrogradation'.

Indigenous starches are often modified chemically, enzymatically, or physically to reform their practical efficiency, including cross-linking or pre-gelatinization, the addition of side groups, and hydrolysis. Modified starches have also appeared to be effective emulsifiers due to their non-polar groups connected to their main branch (Sweedman, Tizzotti, Schäfer, & Gilbert, 2013).

In short, diverse types of starch, such as molecular shape, modified granules, or their natural formulation, may be used to design colloidal delivery systems with various practical attributes (McClements, 2014).

7.4.6.6   Chitosan

Chitosan is another polysaccharide utilized to construct delivery systems. This substance is produced by the deacetylation of chitin, the most abundant compound after cellulose which is also the major substance of fungal cell walls and crustacean's outer skeletons, in the presence of an alkaline (Elieh-Ali-Komi & Hamblin, 2016; Liao et al., 2021). Chitosan has been considered a suitable compound for preparations of nanocarrier systems due to its several properties, including biodegradability, pH sensitivity, easy chemical modification, non-toxicity, and excellent film, gel, and particle forming (Shariatinia, 2019; X. Wang et al., 2014).

7.4.6.7   Cyclodextrin

Cyclodextrins are cyclic oligosaccharides derived from starch and are used for nanoencapsulation of active ingredients. Cyclodextrin's molecular structure consists of an internal lipophilic hole created by a pyranose ring and an external hydrophilic structure. Cyclodextrins are already accepted as 'GRAS' compounds by the United

States Food and Drug Administration, thus can be utilized in the food industry. Nanocapsules prepared by cyclodextrins can be carried out in two forms: solid and liquid, and they also present an acceptable controlled release system (Chaudhari, Singh, Das, & Dubey, 2021).

7.4.6.7 Gums

Various gums are derived from different sources, including microorganisms, animals, and plants. Generally, gums are non-toxic and biodegradable with good physicochemical characteristics, enabling the encapsulation of active components (Padil et al., 2020; Taheri & Jafari, 2019).

One of the most widely used polysaccharides in the food industry is gum arabic, primarily because of its capability as a natural emulsifier in beverage emulsions. This tree exudate is isolated from *Acacia Senegal* and includes numerous high molecular weight fractions (Cui, 2005). Gum arabic has a comparatively dense structure leading to its high water solubility and low solution viscosity compared to alternative gums. Also, above pH three, the charge of the molecule will be negative, which is important while using it in the formation of delivery systems.

Xanthan gum is also a natural extracellular polysaccharide with a high molecular weight, non-toxic, and biocompatible properties, which is produced by *Xanthomonas compestris* bacteria. This gum is typically used to improve the stability of food because of its ability to hydrate in cold water, forming a viscous three-dimensional lattice-like network (Chaudhari et al., 2021).

7.4.6.8 Cellulose

Cellulose is the most abundant polysaccharide in nature, as it is a significant portion of plant structures. Cellulose includes a linear chain of glucose units linked by D-β-(1–4) bonds (Devi, Sarmah, Khatun, & Maji, 2017). Due to its numerous hydrogen bonds and crystalline structure, it has very low solubility in water. Nevertheless, it can be chemically modified in multiple ways to generate products such as ethylcellulose, hydroxyl propyl cellulose, hydroxyl propyl cellulose, and methylcellulose. These are beneficial structural ingredients for biopolymer-based delivery systems (Bao et al., 2020; McClements, 2014; Rostami, Yousefi, Khezerlou, Aman Mohammadi, & Jafari, 2019).

CMC is produced by the chemical attachment of carboxymethyl groups to the main chain of cellulose. This anionic linear polymer, also known as cellulose gum, is commercially sold in two primary forms of calcium or sodium salts. These solutions are sensitive to pH and ionic strength, factors affecting their viscosity (McClements, 2014).

7.4.6.9 Alginates and Carrageenans

Alginate is a polysaccharide derived from certain species of brown seaweed and includes a group of unbranched binary copolymers. Alginate is a desirable compound for encapsulating active components into a gel-based matrix. Moreover, alginate shows a pleasant affiliation to interact with divalent metal ions (such as calcium and magnesium). Thus, gel networks can be easily made in their existence, enhancing their application in the food industry (Chaudhari et al., 2021).

Carrageenan is an anionic polymer extracted from red seaweeds. This sulphated polysaccharide including alternative α-(1–4) and β-(1–3) linked galactose units (McClements, 2014; Nouri et al., 2020).

The final product of drugs encapsulation and their active components prepared by any of the systems mentioned above must be dried afterwards. In the following, an attempt has been made to touch upon spray drying and freeze-drying techniques, which both belong to the atomization methods and are used for transforming liquid suspensions and solutions into more advantageous forms, i.e. powders.

7.5 SPRAY DRYING ENCAPSULATION

Spray drying has been used for many years in industries to produce powdered and solid forms of active components from liquid suspensions and solutions. This technique is economically appropriate for the large-scale production of food ingredients. The required equipment is already being used widely in the food industry to encapsulate various active compounds such as flavour oils, probiotics, colours, peptides, vitamins, bioactive lipids, and proteins. Transforming a liquid form of an active component into powder results in some advantages. It may enhance the shelf life and stability of unstable active ingredients by forming an impermeable solid matrix around the active component. This could delay molecular dispersion processes. Therefore, the formation of the solid matrix will inhibit the interaction of active ingredients with other molecules. Also, this would improve the stability of the active ingredients by inhibiting expulsion or evaporation. The production of powder of an active ingredient from its liquid form can enhance its utilization and handling and lessen storage and transport costs. Lastly, spray drying could encapsulate components sensitive to heat (such as colours, bioactive lipids, probiotics, and flavours). The reason is the fast-drying process, and the latent heat of the evaporated solvent from particles. The temperature will rise relatively low during the process, resulting in spray drying being a mild method (McClements, 2014).

The primary liquid of the ingredients employed in the spray drying process could be in various forms, such as hydrogel particles suspended in the solvent or a simple solution. Examples of such formulations are a protein or carbohydrate dissolved in a solvent, solid-lipid nanoparticles, or a colloidal dispersion like an emulsion droplet. Water is the most commonly used solvent in the food industry, but different solvents can also be utilized for special applications.

The final product of a spray-dried compound contains a complex inner structure, including some large solid particles with smaller colloidal particles within them. These may be agglomerated or coated, forming even more complicated structures later. This capability produces colloidal delivery systems with various practical specifications (McClements, 2014).

Figure 7.4 illustrates the basic principles of the spray drying process. The spray drying method requires various steps. In the beginning, a liquid in which the active component is dissolved (feed fluid) will be into the *atomizer* that can convert the bulk liquid into a haze of tiny drops.

The small drops will then be sprayed into a *drying chamber* where the temperature is kept considerably high so that the solvent, usually water, is evaporated from

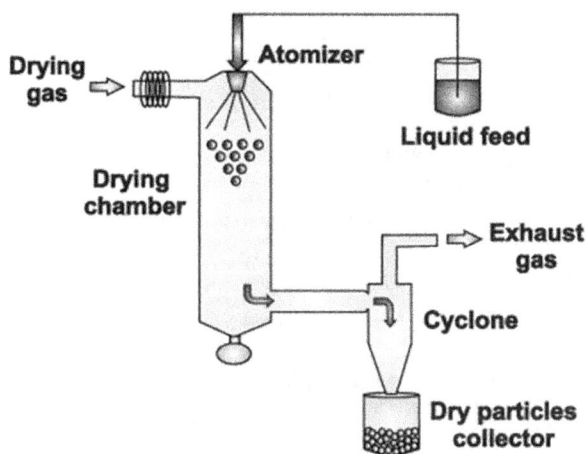

FIGURE 7.4 Basic principle of spray drying method for transforming feed fluids into powders. (From Sosnik and Seremeta (2015) with permissions.)

the *spray drops*. Though the temperature of the heated chamber is far higher than the temperature tolerated by the active ingredients, the endothermic enthalpy change corresponding to evaporation would protect heat-sensitive materials from degradation. The particles of the spray-dried powders are generally between the range of 10 and 100 μm in diameter. Then, the dried particles are assembled using a filter bag or cyclone. After that, the produced particles may undergo additional processing to improve their functional characteristics, storage, handling, or stability by coating or agglomerating (McClements, 2014).

Particles size, retention efficiency, packing density, dispersibility, stability, flowability, and encapsulation efficiency are critical factors in producing a desirable final product/powder.

The composition and formation of the feed liquid are some of the elements affecting the final properties of the produced powder. As mentioned before, the active ingredient is dissolved in a solution which includes the substance which eventually forms the wall material. Water-insoluble active compounds are usually combined with suspension or emulsion, while water-soluble ingredients are dissolved in an aqueous phase. The physical and chemical characteristics of the solid matrix formed at the end of the drying process and the active ingredients are highly affected by the concentration and type of the compounds of materials constructing the walls. They can affect the rheology, permeability, molecular mobility, and dissolution properties of the solid matrix. Thus, choosing an appropriate wall material is a critical step for achieving the desired physicochemical properties needed for specific applications (McClements, 2014).

Also, to obtain a high-quality product, the operating conditions and the spray driers design must be optimized. Due to differences in the quantity, type, and versatility of materials processed, the capital and operating costs, and the nature of the powder produced, different types of spray dryers are available for laboratory, pilot-scale, and industrial use.

One of the main features influencing the efficiency of spray dryers is the gadget used to atomize the samples. This device is either a spray nozzle or a rotary atomizer (McClements, 2014). Various designs of atomizers are commercially available to enhance the productivity of the spray drying process to improve the final product quality. After selecting the proper equipment, operating parameters, including flow rate, outlet, and inlet temperatures, must be optimized. As mentioned previously, it is possible to modify powder characteristics after production to improve their handling or storage. Thus, considerable attention has to be given to the post-drying processes, such as fluid bed methods used to agglomerate or coat the final particles (McClements, 2014; Meiners, 2012).

7.6 FREEZE-DRYING ENCAPSULATION

Freeze-drying or lyophilization is a drying process (Figure 7.5) in which the product is dried by removing its moisture *via* desorption and sublimation methods within the container system. Removing enough water with the usual drying method in a furnace usually requires high temperatures, which are unsuitable for heat-sensitive active materials. High temperatures are not necessary for this process to obtain a sufficiently dry product. Thus, freeze-drying is a major problem in the production of small molecules that are planned for regeneration (Siow, Wan Sia Heng, & Chan, 2016).

Freeze-drying is one of the processes utilized to improve the dissolution of poorly water-soluble drugs and is a procedure known to have the ability to create solid dispersions from amorphous medications. The increase in drug dissolution rate is due to four diverse mechanisms. Primarily, the increased number of free drug molecules available for absorption, the enhanced wetness of the drug due to direct contact with hydrophilic matrix, crystallinity, and finally decrease in particle size, which leads to surface area expansion (Siow et al., 2016).

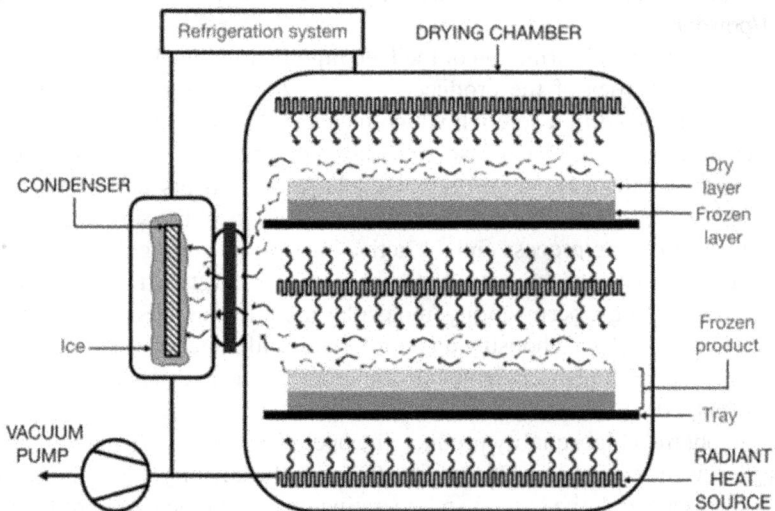

FIGURE 7.5 Basic components of a lyophilization system. (Garcia-Amezquita, Welti-Chanes, Vergara-Balderas, and Bermúdez-Aguirre, 2016.)

Freezing, primary drying, and secondary drying are the three steps of the freeze-drying process. A freeze-dryer typically used in production scale consists of a chamber including several shelves containing trays that control the temperature for drying. This drying chamber is connected to a condenser through a massive valve. One or several vacuum pumps are also connected to both chambers to provide pressure ranging between 4 and 40 Pa during the entire process (Tang & Pikal, 2004).

During the first step, freezing (also known as solidification), the temperature is decreased below the freezing point of water. As the freezing process goes on, the amount of water frozen in the fluid also increases, leading to a boost in the liquid concentration, which also causes the liquid to become more viscous, yielding a crystalline, amorphous, or combined crystalline-amorphous structure (Patel & Pikal, 2011).

After that, heat will be transmitted to the completely frozen aqueous solution through the trays. In this step, water vapour is produced from ice sublimation inside the chamber under conditions below the vapour pressure of water. Then it will be transferred to the condenser. This would generate pores inside the material.

The secondary drying stage involves eliminating the unfrozen water, which is not sublimed during the first step. This water is also known as bound water.

To freeze-dry a compound, considerable attention must be given to its thermophysical properties. For instance, many components of the nanoparticles formulation, such as the nature of the surfactant, the concentration of cryoprotectant, the polymer used for the formation of nanoparticles, or the chemical groups attached to the surface of nanoparticles, have an extreme effect on their resistance to different stresses caused during freeze-drying process. The freeze-drying process tends to generate many stresses, especially during freezing and dehydration, which can destabilize colloidal suspensions (Abdelwahed, Degobert, Stainmesse, & Fessi, 2006).

Bulking agents, cryoprotectants, and buffers are some of the typically added excipients in traditional freeze-drying formulations (Baheti, Kumar, & Bansal, 2016). For instance, polymers and sugars are usually added in traditional formulizations for variant purposes. Mannitol and polyethyleneglycol act as crystallizing bulking excipient, and their crystallization depends on the relative concentration of other components in the formulation. Amorphous bulking agents such as trehalose and polyvinyl pyrrolidone (PVP) are added to the formulations consisting of proteins as stabilizers because proteins tend to form an amorphous structure after the entire process (Siow et al., 2016).

In novel freeze-drying applications, some water-insoluble excipients have also been used, as well as common ones such as microcrystalline cellulose (MCC). These are utilized in the evolution of innovative freeze-dried excipients for multiple unit pellet systems (MUPS) (Siow et al., 2016).

7.7 PREPARATION AND STANDARDIZATION OF PRODUCTS FROM DRIED HERBS, SPICES, AND MEDICINAL PLANTS

Quality is one of the most important issues considered by consumers in herbal drug utilization (Balekundri & Mannur, 2020). Rigorous guidelines are existed to control the quality of pharmaceuticals. These regulations assure the safety of pharmaceutical products. But unfortunately, in the case of herbal products, the law is not strictly enforced, leading to quality reduction. Different procedures for examination and quality control of herbal drugs are followed such as

- analytical appraisal
- microbiological appraisal
- microscopic appraisal
- physical appraisal
- organoleptic appraisal
- chemical appraisal

Organoleptic assessment of drugs refers for evaluating a drug through odour, size, colour, taste, shape, and unique features such as texture, touch, etc.

Microscopic examination is integral within the preliminary identity of herbs, in addition to figuring out small fragments of crude or powdered herbs and foreign matter and adulterants detection.

Chemical parameter evaluations are used to determine specific drugs or to examine their purity. The chemical strategies of appraisal include purification, isolation, and identification of the active substance. Tests like saponification value, and acid value, are beneficial in the examination, e.g. gums (volatile acidity and methoxy determination), balsams (saponification value, ester values, and acid value,), volatile oils (acetyl and ester values), and resins (acid value, sulphated ash). Preliminary phytochemical screening is part of the chemical examination. And additionally, chemical assessments for herbals, chromatography may be accomplished by using HPLC, TLC, GC, UV, HPTLC, Fluorimeter, GC-MS, etc.

Microbiological contaminations may also happen through cross-contamination from employees who are carriers of pathogenic microorganisms during harvest, post-harvest processing, and the production process. Tests include total mold count, coliform count, and viable content, etc.

Physical parameters are sometimes taken into account to examine certain drugs. These include specific gravity, refractive index, moisture content, optical rotation, viscosity, solubility in different solvents, melting point, clarity, pH, friability, disintegration time, flowability, hardness, settling rate, ash values, and sedimentation.

There are diverse techniques to examine the quality and quantity of herbs, such as SFC (super-critical fluid chromatography), HP-TLC (high-performance thin-layer chromatography), HPLC (high-pressure/performance liquid chromatography), LC-MS (liquid chromatography-mass spectroscopy), ICP-MS (inductively coupled plasma-mass spectroscopy), and GC-MS (gas chromatography-mass spectroscopy) (Balekundri & Mannur, 2020).

The type of phytochemical compounds in herbal material depends on the composition, climate, and soil constituents. Thus, different phytochemicals can prevent the integrity of the standardization process. The deforestation increase has positive effects on the increased adulteration of herbal drugs. Determining physical, chemical, and biological properties would help assess herb's purity and corresponds to freshness. World Health Organization (WHO) has to arrange guidelines for the standardization of herbal drugs based on parameters such as ash values, moisture content, flavour, and aroma, microbial contamination, and spectrometric and chromatographic evaluations (Balekundri & Mannur, 2020). Organoleptic tests, physicochemical studies, and pharmacognostic are required for the authentication and standardizations (Deogade & Prasad, 2019). To adulterants obstructions of herbal drugs, the data obtained from microscopic and macroscopic are very efficient. It is also useful for validating standard parameters and for recognizing the secondary metabolites (Shaheen et al., 2018; Steinhoff, 2019).

7.7.1 Thermal Analysis

The techniques such as thermogravimetric analysis (TGA) are used to determine the thermal stability of samples that demonstrate mass and enthalpy variation, high sensitivity, and reproducibility (Guimarães et al., 2018). The thermal method could be utilized to evaluate the purity, thermal stability, and raw material quality. Also, crystal water content, thermal reduction, and absolute water content can also be obtained by thermal analysis (Cuinica & Macêdo, 2018; Liu, Yang, Ma, & Zhang, 2019). The two main parameters in the thermal analysis are time and temperature. In this analysis, samples are examined in isothermal and non-isothermal conditions. To evaluate the interaction between the excipient and active pharmaceutical ingredients (API), the technologies such as DTG (different TGA) and DSC (differential scanning calorimetry) are utilized. For the recognition of thermal behavior of polymer, TG and DSC are used (Toma, Tita, Olah, & Statti, 2018).

7.7.2 High-Performance Liquid Chromatography (HPLC)

This technique is utilized to determine the safety, quality, and efficacy of phytochemical components (Do, Santi, & Reich, 2019). Modern analytical techniques such as high-performance thin-layer chromatography (HPTLC) are used to eliminate obstacles in assessing the quality control aspects of herbal materials (Bhurat, Sanghavi, Nagdev, & Patil, 2018; Chewchida & Vongsak, 2019).

7.7.3 Super-critical Fluid Chromatography (SFC)

Since the chemical solvent causes environmental pollution, SFC is an alternative method (Gitea et al., 2018). This system consists of compressed carbon dioxide and organic solvent. This method requires less quantity of organic solvent and less pressure for the mobile phase compared to liquid chromatography (T.-T. Liu et al., 2019). SFC technique can analyse lipid, alkaloids, flavonoids, carbohydrates, saponin, etc. The advantages of this technique include low time consumption, being eco-friendly, and high sensitivity in the analysis of polar and non-polar compounds (Balekundri & Mannur, 2020).

7.7.4 Inductively Coupled Plasma-Mass Spectroscopy (ICP-MS)

The presence of elements in medicinal plants performs a beneficial role in the human's diet. The elementary composition of herbs depends on parameters such as the geochemical nature of the soil. The elements that are available in the plants are divided into two forms: ionic and non-ionic. Some of them are toxic. Cadmium, lead, and mercury are toxic even in low concentrations (Varhan Oral et al., 2019). ICP-MS is a method to measure the medicinal plant elemental content (Rao, Gowri Naidu, Sarita, Srikanth, & Naga Raju, 2019).

7.7.5 LC-MS

This method is a combination of two methods: mass spectroscopy (MS) and HPLC technique. This technique possesses high sensitivity and resolution. The parameters which can be measured by LC-MS include structure characterization, retention time, and molecular mass (Wu et al., 2018).

7.7.6 GC-MS

The extract of herbal materials can be analysed by the GC-MS technique. Analysis of thermostable volatile compounds is one of the most important analyses that is carried out (Rutkowska, Łozowicka, & Kaczyński, 2018; Taha & Gadalla, 2017).

7.8 DRIED HERBS QUALITY CONTROL

7.8.1 Product-Specific Quality Parameters

Herbs and especially spices have been considered vulnerable and high price commodities. Thus, essential worldwide requirements are set by the USA and European Union (EU). Additionally, a few standards exist in countries responsible for growing herbs and spices. Various tests make up the variety of worldwide requirements and standards:

- Ash level: This evaluation is obtained by burning the organic matter in the product; thus, it shows the product level of impurities and ash.
- Cleanliness: This measurement indicates the amount of extraneous and foreign matters, for example, foreign bodies and insect contamination.
- Volatile oil (V/O) determination: This evaluation helps to display herb or spice has been adulterated or not, such as the foreign materials addition, low-quality materials, or spent herbs.
- Acid insoluble ash (AIA) (or sand content): This is a conventional assurance of the herb's cleanliness.
- Water activity: In the latest years, moisture content has been associated with the herbs' water activity (a_w).
- Moisture content: This is an important measurement since moisture content shows the weight that has been used in pricing.
- Pesticide levels: These parameters are not seen as a significant health problem given the (low) average daily intake of herbs and spices by consumers.
- Microbiological levels: Various techniques for calculating microorganisms in herbs are available.

- Mycotoxin levels: There is a growing concern about mycotoxins, in particular, aflatoxin and ochratoxin A in the herbal medicine manufacturing industry.
- Mesh size: This technique is performed *via* using standard sieves that are frequently characterized in 'micron sizes', and aids the dispersion of flavour. Also, standard specification necessities could be a 95% pass on a specified size of the sieve.

Bulk density: This is an important measure, particularly in filling retail containers of herbs and spices. The herb or spices must be sifted or ground to give a certain density so that retail units not only appear satisfactorily full but also comply with the declared weight. Densities may be measured packed down, e.g., after tapping the product so that it assumes a minimum density, or untapped

- ASTA colour values: This measurement displays the extractable colour of products of the Capsicum genus, and it has been used as a paprika quality indicator. For the extraction acetone (16-hour ambient extraction) is used, observation through spectrophotometric analysis (against a standard at 460 nm) is used.
- Curcumin content: This is a specific measurement of the extracted turmeric colour. It has been performed *via* reflux extraction in acetone accompanied by a spectrophotometer at 415–425 nm.
- Scoville heat units: The Scoville heat unit is a capsaicin content measurement of Capsicum species. The test includes extraction of the capsaicin in alcohol and tasting of successively more potent dilutions in sugar syrup till the chili heat is detected.
- Non-permitted hues: These hues can be added illegally into spices such as fennel, chilies, paprika, saffron, cassia, and star anise to increase their value and physical appearance. Non-permitted colours are assessed by liquid chromatography or HPLC with a diode array detector (DAD) and tandem mass spectrometry (LC/MS/MS) depending on the restriction of detection (LOD) required (Kuruppacharil V Peter, 2006).

7.8.2 Packaging and Labelling

To maintain processed herbal quality, they must be packaged as quickly as possible. Packaging must prevent herbal medicines deterioration and be resistant to contamination and pest infestations. The maximal maintenance of unpacked herbal medicines should be established when applicable. Continuous in-process QC measurements must be implemented to discard substandard materials, contaminants, and foreign matter during the final packaging stages. Herbal products should be packaged in clean, breathable bags, dry boxes, or different packing containers according to the standard operating procedures (SOPs) and should follow the national regulations of the end-consumer countries and the manufacturer. Materials used for packaging must be clean, non-polluted, undamaged, and dry, and should maintain the processed herbal quality requirements. Fragile natural substances have to be packaged in inflexible containers. Wherever possible, the packaging used should be agreed upon between the provider and the consumer. A label affixed to the packaging must include the following data:

- The official name of the herb, herbal preparation, herbal material, or herbal dosage form.
- The accepted scientific herb name.
- Processing of the processed herb date.
- Brand name of the herb, herbal preparation, herbal material, or herbal dosage form.
- Method of processing.
- Active ingredient potency or strength.
- Names and addresses of the herbal preparations' processor or herbal materials, herbal dosage forms (finished herbal products), importer, or distributor.
- Net amount in the container in terms of weight and unit number.
- Excipients lists.
- Suggested storage conditions.
- Date of expiration.
- Batch number.

The label should also indicate quality approval data and national or regional labelling requirements.

Records of batch packaging should be displayed and must contain product name, batch number, assignment number, place of origin, weight, and date. The records should be kept as required by national and/or regional authorities or retained for a period of 3 years.

7.9 STORAGE AND TRANSPORTATION

Before intake, all manufactured herbal medicines should be stored and preserved properly. The vicinity should be free from contaminations such as insects, microorganisms, and pests. Proper packaging must be used to give the best safety against physical harm to the processed materials; and from exposure to heat, moisture, insects, light, and animals. Rejected samples should be stored in a separate specified designated quarantined area, vividly labelled, and with a specified handling period.

Controlled or poisonous herbal materials or arrangements should be checked, labeled, and kept in line with the government's law. Storage regions should have enough capacity to permit orderly storage of diverse processed herbal preparations, herbal materials, or herbal dosage forms with the proper separation and segregation. They must be monitored, controlled, and recorded where suitable to ensure certain proper storage situations and follow with the "first-in and first-out" principle.

Conveyances for transporting processed herbal medicines from the processing region to the storage region should be appropriate, clean, and suitably ventilated to prevent condensation and maintain appropriate airflow.

Skilled or trained personnel should facilitate pest infest control in conveyances and storage regions. Only registered and authorized chemical agents should be used.

All fumigation and dates of application should be kept. For pest management, when freezing or saturated steam is used, the humidity of the stored herbal medicines should be checked.

7.10   EQUIPMENT

According to WHO, all equipment consisting of utensils and tools used in the herbal manufacturing methods should be made from substances that do not transmit odour, taste, or toxic substances. They should be non-absorbent and corrosion-resistant. The materials also should be able to withstand repeated cleaning and disinfection. The use of wooden and different types of materials that cannot be effectively cleaned and disinfected must be avoided, except when their use would not be a source of contamination. Also, the use of metals that cause corrosion must be avoided.

All equipment should be designed and built to allow easy and thorough cleaning and disinfection, additionally to prevent hygiene hazards. Where practicable, they ought to be reachable for visible inspection. Containers for unusable substances or waste must be leak-proof, built of metal, or other appropriate impervious substances. They must be easy to clean or disposable and close securely. Finally, all refrigerated areas should be equipped with temperature measurement and recording devices (WHO, 2018).

7.11   CONCLUDING REMARKS

Herbal products are part of the conventional medical system and have been extensively used all around the globe since ancient times. Recently, scientists have switched their consciousness to designing novel drug delivery systems for herbs. Various strategies are required to manipulate the quality of such products. However, studies in this region are still in the exploratory stage. A range of plant components such as tannins, flavonoids, etc., confirmed improved healing impact at comparable or less dose when incorporated into novel drug delivery systems. In contrast with conventional plant extracts, they will offer more secure use and effective remedy. However, these formulations have immense potential to deliver greater advantages for society and humans by offering means of well-being. It is expected that commercial production and widespread utilization of these novel products will be available in the market in the near future.

REFERENCES

Abdelwahed, W., Degobert, G., Stainmesse, S., & Fessi, H. (2006). Freeze-drying of nanoparticles: Formulation, process and storage considerations. *Advanced Drug Delivery Reviews*, 58(15), 1688–1713. https://doi.org/10.1016/j.addr.2006.09.017

Akhilesh, D., Faishal, G., & Kamath, J. (2012). Comparative study of carriers used in proniosomes. *International Journal of Pharmaceutical Chemistry and Analysis*, 3, 6–12.

Baheti, A., Kumar, L., & Bansal, A. K. (2016). Excipients used in lyophilization of small molecules. *Journal of Excipients and Food Chemicals*, 1(1), 1135.

Balekundri, A., & Mannur, V. (2020). Quality control of the traditional herbs and herbal products: A review. *Future Journal of Pharmaceutical Sciences*, 6(1), 1–9.

Bao, X., Dong, F., Yu, Y., Wang, Q., Wang, P., Fan, X., & Yuan, J. (2020). Green modification of cellulose-based natural materials by HRP-initiated controlled "graft from" polymerization. *International Journal of Biological Macromolecules, 164*, 1237–1245. https://doi.org/10.1016/j.ijbiomac.2020.07.248

Berkulin, W., & Theissing, K.-H. (2003). *Process for preparing dry extracts.* Google Patents.

Bhurat, M. R., Sanghavi, R. S., Nagdev, S. A., & Patil, D. (2018). HPTLC pingerprinting and quantification of SHATAVARIN IV in extract and polyherbal formulations. *World Journal of Pharmaceutical Research, 7*(10), 442–451.

Boussetta, N., & Vorobiev, E. (2014). Extraction of valuable biocompounds assisted by high voltage electrical discharges: A review. *Comptes Rendus Chimie, 17*(3), 197–203.

Chaudhari, A. K., Singh, V. K., Das, S., & Dubey, N. K. (2021). Nanoencapsulation of essential oils and their bioactive constituents: A novel strategy to control mycotoxin contamination in food system. *Food and Chemical Toxicology, 149*, 112019.

Chemat, F., & Strube, J. (2015). *Green Extraction of Natural Products: Theory and Practice.* John Wiley & Sons.

Chewchida, S., & Vongsak, B. (2019). Simultaneous HPTLC quantification of three caffeoylquinic acids in *Pluchea indica* leaves and their commercial products in Thailand. *Revista Brasileira de Farmacognosia, 29*, 177–181.

Cui, S. W. (2005). *Food Carbohydrates: Chemistry, Physical Properties, and Applications.* CRC Press.

Cuinica, L. G., & Macêdo, R. O. (2018). Thermoanalytical characterization of plant drug and extract of *Urtica dioica* L. and kinetic parameters analysis. *Journal of Thermal Analysis and Calorimetry, 133*(1), 591–602.

Deogade, M. S., & Prasad, K. (2019). Standardization of wild Krushnatulasi (*Ocimum tenuiflorum* Linn) leaf. *International Journal of Ayurvedic Medicine, 10*(1), 52–61.

Devi, N., Sarmah, M., Khatun, B., & Maji, T. K. (2017). Encapsulation of active ingredients in polysaccharide–protein complex coacervates. *Advances in Colloid and Interface Science, 239*, 136–145. https://doi.org/10.1016/j.cis.2016.05.009

Do, T., Santi, I., & Reich, E. (2019). A harmonized HPTLC method for identification of various caffeine containing herbal drugs, extracts, and products, and quantitative estimation of their caffeine content. *Journal of Liquid Chromatography & Related Technologies, 42*(9–10), 274–281.

Elieh-Ali-Komi, D., & Hamblin, M. R. (2016). Chitin and chitosan: Production and application of versatile biomedical nanomaterials. *International Journal of Advanced Research, 4*(3), 411.

Eltawil, M. A., Azam, M. M., & Alghannam, A. O. (2018). Energy analysis of hybrid solar tunnel dryer with PV system and solar collector for drying mint (MenthaViridis). *Journal of Cleaner Production, 181*, 352–364.

Erbas, S., & Baydar, H. (2016). Variation in scent compounds of oil-bearing rose (*Rosa damascena* Mill.) produced by headspace solid phase microextraction, hydrodistillation and solvent extraction. *Records of Natural Products, 10*(5), 555.

Fang, J., Wang, Z., Wang, P., & Wang, M. (2020). Extraction, structure and bioactivities of the polysaccharides from Ginkgo biloba: A review. *International Journal of Biological Macromolecules, 162*, 1897–1905.

Garcia-Amezquita, L. E., Welti-Chanes, J., Vergara-Balderas, F. T., & Bermúdez-Aguirre, D. (2016). Freeze-drying: The basic process. In B. Caballero, P. M. Finglas, & F. Toldrá (Eds.), *Encyclopedia of Food and Health* (pp. 104–109). Academic Press.

Giacometti, J., Kovačević, D. B., Putnik, P., Gabrić, D., Bilušić, T., Krešić, G., … Barbosa-Cánovas, G. (2018). Extraction of bioactive compounds and essential oils from mediterranean herbs by conventional and green innovative techniques: A review. *Food Research International, 113*, 245–262.

Gitea, D., Vicas, S., Gitea, M. A., Nemeth, S., Tit, D. M., Pasca, B., & Iovan, C. (2018). HPLC screening of bioactives compounds and antioxidant capacity of different hypericum species. *Revista de Chimie*, *69*(2), 305–309.

Gouda, M., Bekhit, A. E.-D., Tang, Y., Huang, Y., Huang, L., He, Y., & Li, X. (2021). Recent innovations of ultrasound green technology in herbal phytochemistry: A review. *Ultrasonics Sonochemistry*, *73*, 105538.

Goyal, A., Kumar, S., Nagpal, M., Singh, I., & Arora, S. (2011). Potential of novel drug delivery systems for herbal drugs. *Indian Journal of Pharmaceutical Education and Research*, *45*(3), 225–235.

Guimarães, G. P., Santos, R. L., Brandão, D. O., Cartaxo-Furtado, N. A. D. O., Cavalcanti, A. L. D. M., & Macedo, R. O. (2018). Thermoanalytical characterization of herbal drugs from *Poincianella pyramidalis* in different particle sizes. *Journal of Thermal Analysis and Calorimetry*, *131*(1), 661–670.

Hatano, K.-I., Miyakawa, T., Sawano, Y., & Tanokura, M. (2011). Antifungal and lipid transfer proteins from Ginkgo (*Ginkgo biloba*) Seeds. In *Nuts and Seeds in Health and Disease Prevention* (pp. 527–534). Elsevier.

Jacob, S., Kasthurirengan, S., Karunanithi, R., & Behera, U. (2000). Development of pilot plant for cryogrinding of spices: A method for quality improvement. *Advances in Cryogenic Engineering*, *45*, 1731–1738.

Jadhav, A. I., Wadhave, A. A., Arsul, V. A., & Sawarkar, H. (2014). Phytosomes: A novel approach in herbal drug delivery system. *International Journal of Pharmaceutics and Drug Analysis*, *2*, 478–486.

Leal, P. F., Almeida, T. S., Prado, G. H., Prado, J. M., & Meireles, M. A. A. (2011). Extraction kinetics and anethole content of fennel (*Foeniculum vulgare*) and anise seed (*Pimpinella anisum*) extracts obtained by soxhlet, ultrasound, percolation, centrifugation, and steam distillation. *Separation Science and Technology*, *46*(11), 1848–1856.

Lee, M.-K. (2020). Liposomes for enhanced bioavailability of water-insoluble drugs: *In vivo* evidence and recent approaches. *Pharmaceutics*, *12*(3), 264.

Liao, W., Badri, W., Dumas, E., Ghnimi, S., Elaissari, A., Saurel, R., & Gharsallaoui, A. (2021). Nanoencapsulation of essential oils as natural food antimicrobial agents: An overview. *Applied Sciences*, *11*(13), 5778.

Liu, T.-T., Cheong, L.-Z., Man, Q.-Q., Zheng, X., Zhang, J., & Song, S. (2019). Simultaneous profiling of vitamin D metabolites in serum by supercritical fluid chromatography-tandem mass spectrometry (SFC-MS/MS). *Journal of Chromatography B*, *1120*, 16–23.

Liu, Y., Yang, L., Ma, C., & Zhang, Y. (2019). Thermal behavior of sweet potato starch by non-isothermal thermogravimetric analysis. *Materials*, *12*(5), 699.

McClements, D. J. (2014). *Nanoparticle and Microparticle-Based Delivery Systems: Encapsulation, Protection and Release of Active Compounds*. CRC Press.

Meiners, J. (2012). Fluid bed microencapsulation and other coating methods for food ingredient and nutraceutical bioactive compounds. In *Encapsulation Technologies and Delivery Systems for Food Ingredients and Nutraceuticals* (pp. 151–176). Elsevier.

Mukerjee, A., & Vishwanatha, J. K. (2009). Formulation, characterization and evaluation of curcumin-loaded PLGA nanospheres for cancer therapy. *Anticancer Research*, *29*(10), 3867–3875.

Negbi, M., Dagan, B., Dror, A., & Basker, D. (1989). Growth, flowering, vegetative reproduction, and dormancy in the saffron crocus (*Crocus sativus* L.). *Israel Journal of Plant Sciences*, *38*(2–3), 95–113.

Nouri, A., Tavakkoli Yaraki, M., Lajevardi, A., Rahimi, T., Tanzifi, M., & Ghorbanpour, M. (2020). An investigation of the role of fabrication process in the physicochemical properties of κ-carrageenan-based films incorporated with Zataria multiflora extract and nanoclay. *Food Packaging and Shelf Life*, *23*, 100435. https://doi.org/10.1016/j.fpsl.2019.100435

Padil, V. V. T., Cheong, J. Y., Kp, A., Makvandi, P., Zare, E. N., Torres-Mendieta, R., … Varma, R. S. (2020). Electrospun fibers based on carbohydrate gum polymers and their multifaceted applications. *Carbohydrate Polymers*, *247*, 116705. https://doi. org/10.1016/j.carbpol.2020.116705

Parab, A., Salgaonkar, K., Padwekar, O., & Purohit, S. (2020). Extraction and formulation of perfume from lemon grass. *International Journal of Environmental & Agriculture Research*, *6*(12), 26–30.

Parniakov, O., Barba, F. J., Grimi, N., Lebovka, N., & Vorobiev, E. (2016). Extraction assisted by pulsed electric energy as a potential tool for green and sustainable recovery of nutritionally valuable compounds from mango peels. *Food Chemistry*, *192*, 842–848.

Patel, S. M., & Pikal, M. J. (2011). Emerging freeze-drying process development and scale-up issues. *Aaps Pharmscitech*, *12*(1), 372–378.

Pérez, S., & Bertoft, E. (2010). The molecular structures of starch components and their contribution to the architecture of starch granules: A comprehensive review. *Starch-Stärke*, *62*(8), 389–420.

Peter, K. V. (2006). *Handbook of Herbs and Spices* (volume 3). Woodhead Publishing.

Peter, K. V., & Babu, K. N. (2012). Introduction to herbs and spices: medicinal uses and sustainable production. In K. V. Peter (Ed.), *Handbook of Herbs and Spices* (Second Edition) (pp. 1–16). Woodhead Publishing.

Phillips, G., & Williams, P. (2011). Introduction to food proteins. In G. O. Phillips and P. A. Williams (Eds.), *Handbook of Food Proteins* (pp. 1–12). Woodhead Publishing Limited.

Puertolas, E., Koubaa, M., & Barba, F. J. (2016). An overview of the impact of electrotechnologies for the recovery of oil and high-value compounds from vegetable oil industry: Energy and economic cost implications. *Food Research International*, *80*, 19–26.

Pushpangadan, P., & George, V. (2012). Basil. In K. V. Peter (Ed.), *Handbook of Herbs and Spices* (Second Edition) (pp. 55–72). Woodhead Publishing.

Pushpangadan, P., Rajasekharan, S., & Biju, S. (1993). Tulasi. Tropical Botanic Garden and Research Institute, Thiruvananthapuram, 1–2.

Rambaran, T. (2020). Nanopolyphenols: A review of their encapsulation and anti-diabetic effects. *SN Applied Sciences*, *2*. https://doi.org/10.1007/s42452-020-3110-8

Rao, J., Gowri Naidu, B., Sarita, P., Srikanth, S., & Naga Raju, G. (2019). Elemental analysis of *Pterocarpus santalinus* by PIXE and ICP-MS: Chemometric approach. *Journal of Radioanalytical and Nuclear Chemistry*, *322*(1), 129–137.

Rostami, M., Yousefi, M., Khezerlou, A., Aman Mohammadi, M., & Jafari, S. M. (2019). Application of different biopolymers for nanoencapsulation of antioxidants via electrohydrodynamic processes. *Food Hydrocolloids*, *97*, 105170. https://doi.org/10.1016/j. foodhyd.2019.06.015

Rutkowska, E., Łozowicka, B., & Kaczyński, P. (2018). Modification of multiresidue QuEChERS protocol to minimize matrix effect and improve recoveries for determination of pesticide residues in dried herbs followed by GC-MS/MS. *Food Analytical Methods*, *11*(3), 709–724.

Ruttarattanamongkol, K., Siebenhandl-Ehn, S., Schreiner, M., & Petrasch, A. M. (2014). Pilot-scale supercritical carbon dioxide extraction, physico-chemical properties and profile characterization of *Moringa oleifera* seed oil in comparison with conventional extraction methods. *Industrial Crops and Products*, *58*, 68–77.

Sampathu, S., Shivashankar, S., Lewis, Y., & Wood, A. (1984). Saffron (*Crocus sativus* Linn.)—Cultivation, processing, chemistry and standardization. *Critical Reviews in Food Science & Nutrition*, *20*(2), 123–157.

Sankhyan, A., & Pawar, P. (2012). Recent trends in niosome as vesicular drug delivery system. *Journal of Applied Pharmaceutical Science*, *2*(6), 20–32.

Saraf, S. (2010). Applications of novel drug delivery system for herbal formulations. *Fitoterapia*, *81*(7), 680–689.

Sarangi, M. K., & Padhi, S. (2018). Novel herbal drug delivery system: An overview. *Archives of Medicine and Health Sciences, 6*(1), 171.

Saxena, S., Barnwal, P., Balasubramanian, S., Yadav, D., Lal, G., & Singh, K. (2018). Cryogenic grinding for better aroma retention and improved quality of Indian spices and herbs: A review. *Journal of Food Process Engineering, 41*(6), e12826.

Selvamuthukumaran, M. (2019). *Handbook on Spray Drying Applications for Food Industries.* CRC Press.

Selvamuthukumaran, M., Tranchant, C., & Shi, J. (2019). Spraying drying concept, application and its recent advances in food processing. In M. Selvamuthukumaran (Ed.), *Handbook on Spray Drying Applications for Food Industries* (pp. 1–29). CRC Press.

Shaheen, N., Imam, S., Abidi, S., Sultan, R. A., Azhar, I., & Mahmood, Z. A. (2018). Comparative pharmacognostic evaluation and standardization of *Capsicum annuum* L. (red chili). *International Journal of Pharmaceutical Sciences and Research, 9*(7), 2807–2817.

Shariatinia, Z. (2019). Pharmaceutical applications of chitosan. *Advances in Colloid and Interface Science, 263*, 131–194. https://doi.org/10.1016/j.cis.2018.11.008

Singh, B., Kaur, P., Singh, R., & Ahuja, P. (2008). Biology and chemistry of *Ginkgo biloba. Fitoterapia, 79*(6), 401–418.

Siow, C. R. S., Wan Sia Heng, P., & Chan, L. W. (2016). Application of freeze-drying in the development of oral drug delivery systems. *Expert Opinion on Drug Delivery, 13*(11), 1595–1608.

Siraj, F. M., SathishKumar, N., Kim, Y. J., Kim, S. Y., & Yang, D. C. (2015). Ginsenoside F2 possesses anti-obesity activity via binding with PPARγ and inhibiting adipocyte differentiation in the 3T3-L1 cell line. *Journal of Enzyme Inhibition and Medicinal Chemistry, 30*(1), 9–14.

Sosnik, A., & Seremeta, K. P. (2015). Advantages and challenges of the spray-drying technology for the production of pure drug particles and drug-loaded polymeric carriers. *Advances in Colloid and Interface Science, 223*, 40–54.

Souza, C., & Oliveira, W. (2004). Dried extracts of *Bauhinia forficata* Link obtained by spray drying: Effect of operating parameters on thermal degradation of active compounds and on physical properties of the product. *Proc. of the XV Braz. Cong. of Chem. Eng. (CD-Rom).*

Steinhoff, B. (2019). Quality of herbal medicinal products: State of the art of purity assessment. *Phytomedicine, 60*, 153003.

Sweedman, M. C., Tizzotti, M. J., Schäfer, C., & Gilbert, R. G. (2013). Structure and physicochemical properties of octenyl succinic anhydride modified starches: A review. *Carbohydrate Polymers, 92*(1), 905–920.

Taha, S. M., & Gadalla, S. A. (2017). Development of an efficient method for multi residue analysis of 160 pesticides in herbal plant by ethyl acetate hexane mixture with direct injection to GC-MS/MS. *Talanta, 174*, 767–779.

Taheri, A., & Jafari, S. M. (2019). Gum-based nanocarriers for the protection and delivery of food bioactive compounds. *Advances in Colloid and Interface Science, 269*, 277–295. https://doi.org/10.1016/j.cis.2019.04.009

Tainter, D. R., & Grenis, A. T. (1993). *Spices and Seasonings: A Food Technology Handbook.* Wiley-VCH Verlag GmbH.

Tang, X. C., & Pikal, M. J. (2004). Design of freeze-drying processes for pharmaceuticals: Practical advice. *Pharmaceutical Research, 21*(2), 191–200.

Toma, C.-C., Tita, B., Olah, N.-K., & Statti, G. (2018). Investigation of thermal behavior of *Nigellae sativae* semen from different types of extracts. *Studia Universitatis Babes Bolyai Chemia, 63*(2), 157–164.

Uhrich, K. E., Cannizzaro, S. M., Langer, R. S., & Shakesheff, K. M. (1999). Polymeric systems for controlled drug release. *Chemical Reviews-Columbus, 99*(11), 3181–3198.

Varhan Oral, E., Tokul-Ölmez, Ö., Yener, İ., Firat, M., Tunay, Z., Terzioğlu, P., … Ertaş, A. (2019). Trace elemental analysis of allium species by inductively coupled plasma-mass spectrometry (ICP-MS) with multivariate chemometrics. *Analytical Letters*, *52*(2), 320–336.

Venskutonis, P. R. (2002). Harvesting and post-harvest handling in the genus Thymus. In *Thyme* (pp. 211–237). CRC Press.

Wang, T., Jin, Y., Yang, W., Zhang, L., Jin, X., Liu, X., … Li, X. (2017). Necroptosis in cancer: An angel or a demon? *Tumor Biology*, *39*(6), 1010428317711539.

Wang, X., Chen, Y., Dahmani, F. Z., Yin, L., Zhou, J., & Yao, J. (2014). Amphiphilic carboxy-methyl chitosan-quercetin conjugate with P-gp inhibitory properties for oral delivery of paclitaxel. *Biomaterials*, *35*(26), 7654–7665.

WHO. (2018). WHO guidelines on good herbal processing practices for herbal medicines. *WHO Technical Report Series* (1010).

Wu, W., Jiao, C., Li, H., Ma, Y., Jiao, L., & Liu, S. (2018). LC-MS based metabolic and meta-bonomic studies of Panax ginseng. *Phytochemical Analysis*, *29*(4), 331–340.

Yan, L.-G., He, L., & Xi, J. (2017). High intensity pulsed electric field as an innovative tech-nique for extraction of bioactive compounds—A review. *Critical Reviews in Food Science and Nutrition*, *57*(13), 2877–2888.

8 Emerging and Advanced Drying Technologies for Herbs, Spices, and Medicinal Plants

Klaudia Masztalerz, Jacek Łyczko, and Krzysztof Lech
Wroclaw University of Environmental and Life Sciences

CONTENTS

8.1 INTRODUCTION

Herbs, spices, and medicinal plants (HSMPs) are a broad group of strongly related materials, which are used for different purposes. Firstly, 'herbs' and 'spices' are terms commonly used as synonyms, especially for culinary needs (Raghavan 2006). Nevertheless, this figure of speaking is not scientifically acceptable. According to the Cambridge dictionary, 'spice' is a much narrower term than 'herb', since the very first one may be considered as plant-originated product used for food flavoring, while the following one is particularly a plant, which leaves are used for flavoring and/or for medicine formulations. Therefore, here one may introduce another term – medicinal plant (commonly medicinal and aromatic plant) – which is defined as the

DOI: 10.1201/9781003269250-8

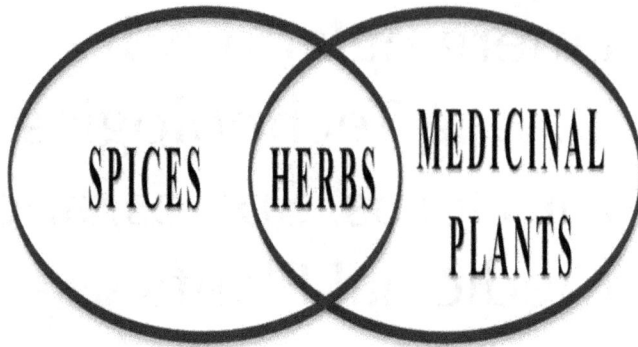

FIGURE 8.1   Relation between herbs, spices, and medicinal plants.

plant material sufficient for disease prevention and treatment (Öztekin and Martinov 2014). In light of that, the relation among those three terms is presented in Figure 8.1.

Drying is a technological process, which aims to remove water from the raw material in order to reduce or eliminate the enzymatic activity and possibility of microorganism development. The general purpose of moisture content decrease is prolonging the dried product shelf life and, additionally, to facilitate the transportation chain procedures and cost efficiency (Öztekin and Martinov 2014). Since the main goal is the water removal, the HSMPs structure (on the cellular level), size, and moisture content has a significant impact on the drying method selection. Moreover, the presence of particular phytochemicals has an influence on that selection. The brief characteristics of commonly used HSMPs, their valuable parts, phytochemicals, and applications are given in Table 8.1.

It can be seen that the diversity of parts of the plant that are utilized along with potential applications is extremely high, which makes it necessary to carefully consider the influence of potential drying methods. Depending on the part of the plant that needs to be preserved, different approaches to drying might be necessary. Since maintaining high quality of HSMPs after drying is crucial, factors affecting the drying of HSMPs are critical in choosing appropriate drying methods. Therefore, their brief characteristic is provided in this chapter. There are several drying methods that have been commonly used for herbs, spices, and medicinal plants. The most popular are solar drying, hot air drying, and vacuum and microwave drying (Jin et al. 2020). However, most of these methods present some significant disadvantages that affect the quality of dried materials and/or require high production costs. Cost effectiveness is a very important aspect of every production process. When considering advanced and emerging drying methods in laboratory-scale investment costs is not really relatable to industry conditions, therefore, it will be neglected in this chapter. Instead, operational costs, which in the case of drying of HSMPs, depend mostly on the type of energy used for drying, duration of the process, energy source, as well as its environmental impact will be evaluated. Therefore, new and advanced drying technologies need to be introduced to ensure both consumers' and producers' demands and will be discussed further. Moreover, drying of HSMPs using traditional and popular methods often leads to poor quality or presents other disadvantages. Therefore, the

TABLE 8.1

Characterization of Commonly Used Herbs, Spices, and Medicinal Plants

Systematic Name	Common Name	Utilized Part	Phytochemicals	Application	References
Zingiber officinale Rosc.	Ginger	Rhizome	Gingerols, shogaols, zingerone, α-zingiberene	Seasoning, beverages production, gastrointestinal treatment, appetite improvement, cold treatment	An et al. (2019) and Zhang et al. (2021b)
Valeriana officinalis L.	Valerians	Roots	Valerenic acid, iridoids, borneol	Gastrointestinal treatment, sedation, sleep quality improvement, anxiety treatment	Chen et al. (2015) and Patočka and Jákl (2010)
Cinnamomum verum J. Presl	Cinnamon	Barks	Cinnamaldehyde, eugenol, caryophyllene, cinnamic acid, cinnamyl acetate	Flavoring, antioxidant activities, antimicrobial activities, anti-inflammatory activity, seasoning	Singh et al. (2021)
Allium sativum L.	Garlic	Bulbs	Alliin, allicin, ajoens, diallyl sulfide, diallyl disulfide, diallyl trisulfide	Antimicrobial, activities, anticancerogenic activities, antyhypertensive activities, food flavoring	El-Saber Batiha et al. (2020)
Mentha x piperita L.	Mint	Leaves	Menthol, mentone, piperitone, isomenthone, neoisomenthone, eriocitirin	Gastrointesitnal treatment, flu treatment, antimicrobial activities, food flavoring, beverages production, hypertension treatment	Mahendran and Rahman (2020)
Artemisia absinthum L.	Wormwood	Herb (leaves and steams)	Thujones, eucalyptol, chamazulene, *cis*-chrysanthenol	Immunomodulatory activities, hepatoprotective, antimicrobial activities, neuroprotective activities, procognitive treatments	Nguyen and Németh (2016) and Szopa et al. (2020)
Coriandrum sativum L.	Coriander	Seeds and fruits	Linalool, geranyl acetate, petroselinic acid, linoleic acid	Seasoning, antidiabetic activities, anti-inflammatory activities, antioxidant activities	Laribi et al. (2015) and Łyczko et al. (2021)
Lavandula angustifolia Mill.	True lavender	Flowers	Linalool, linalyl acetate, caryophyllene, lavandulyl acetate	Sedation, sleep quality improvement, anxiety treatment, aromatization	Łyczko et al. (2019a) and Smigielski et al. (2018)

needs for new and advanced drying technologies, along with their characteristics, are provided. Finally, enhanced and emerging drying methods are summarized and discussed in detail, and future developments in drying of HSMPs are presented.

8.2 FACTORS AFFECTING THE QUALITY OF HERBS, SPICES, AND MEDICINAL PLANTS DURING DRYING

Drying (dehydration) is a multifaceted technological process, which is one of the most significant among HSMPs postharvest treatments due to their future quality. Intuitively, one would say that the most significant influence on HSMPs-based dried products has the parameters of chosen drying techniques; however, this statement is correct only when it is considered in light of the object (plant material) properties. Regarding the plant material properties, the issue is less complex, since considered should be the plant tissue design and HSMPs valuable phytochemicals behavior during the drying process. Unfortunately, the other aspect – drying techniques parameters – is much more complicated, because each method has its own, individual characteristics (Chua et al. 2019a).

The term 'quality', regarding HSMPs, absolutely has to be specified. There is no possibility to create the universal term, since the guaranteed (expected) quality strongly depends on the product user and/or consumer needs. Therefore, the quality should be considered in light of the aspects presented in Figure 8.2.

The most important factor, which influences the dried product quality, is the raw material. This issue consists of the plant material structure (including tissue level), fragmentation, and HSMPs chemotype (Husnu Can Baser and Buchbauer 2016). Regarding the very first point – plant tissue – the research proves that due to the differences in some macroelements (e.g. nitrogen), the plant organs present diverse permeability (Reich et al. 2008), which significantly affects the drying process. Strongly corresponding to the plant material structure is the fragmentation of the raw material before drying. With no doubts, the level of fragmentation, or grinding, if possible, of the material would result in more efficient surface for water removal. Additionally, related to the aspect of material fragmentation and structure is the location of valuable phytochemicals. For instance, EOs pretty often are stored in plant glandular

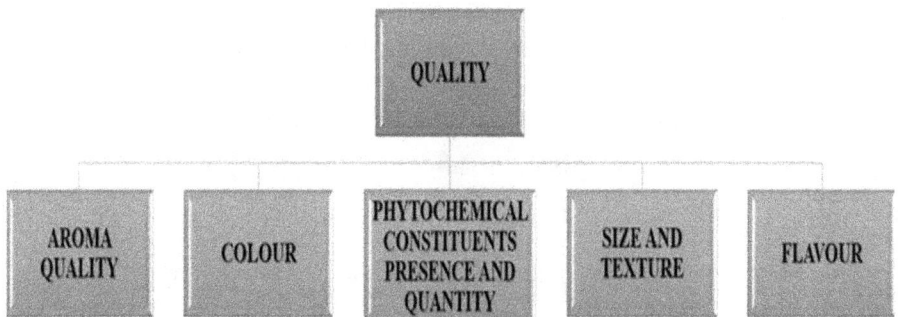

FIGURE 8.2 Main quality characteristics of herbs.

trichomes, which are placed on the plant surface (Glas et al. 2012); thus their physical (regarding their volatility) loss during drying may be highly promoted. Meanwhile, the group of flavonoids, which fulfill the role of dye and are located deeper in plant cells, will be much more robust; however, it still would be vulnerable to residual enzymatic reactions activated during dehydration. Therefore, for EOs yield preservation, shortening the time of drying may be the good solution, while for protecting flavonoids process, temperature decrease will be better.

Finally, the plant chemotype has an influence on the dried product, since the distribution of its phytochemicals has a significant effect on the quality of obtained products and should be strongly confronted with the drying technology choice. For instance, the study performed on *Cannabis sativa* chemotypes (Wanas et al. 2020) had shown that the diverse contribution of volatile organic constituents (VOCs) of essential oils results in different drying procedure performance. As presented in Table 8.2, for major constituents (≥5.0% in at least one chemotype) of *C. sativa* used in the study, there is no possibility to predict the behavior of all plant materials on the basis of the experiment conducted just for one chemotype.

The brief look into Table 8.2 excerpt shows that the behavior of particular VOCs is strongly dependent on their relations to other phytochemicals. The finest example may be the caryophyllene oxide, which in the case of chemotype A it had increased approximately three times, in the case of chemotype B basically it did not change, and for chemotype C it has increased again, however approximately 2.3 times. This issue may be a great risk for adopting a reliable drying protocol, even for regularly used raw material, since the chemotype identification may be obtained only by

TABLE 8.2
An Excerpt from the Volatile Organic Constituents of Essential Oils Obtained from Different Cannabis sativa Chemotypes (Wanas et al. 2020)

Volatile Organic Constituent	Chemotype A Fresh	Chemotype A Dried	Chemotype B Fresh	Chemotype B Dried	Chemotype C Fresh	Chemotype C Dried
				[%]		
α-Pinene	2.2	4.0	15.4	15.6	3.6	6.0
β-Pinene	1.5	2.1	7.8	7.5	2.3	3.5
Myrcene	27.5	17.1	7.9	9.8	10.3	6.9
Limonene	14.0	17.0	33.0	27.1	1.2	1.9
Terpinolene	2.7	tr	–	0.3	9.4	1.7
β-Caryophyllene	7.6	6.2	6.6	9.0	16.5	20.7
α-Humulene	2.6	2.9	3.3	3.8	5.1	6.1
α-Bulnesene	5.5	4.6	1.4	1.5	6.6	8.2
Selina-3,7(11)-diene	4.1	3.7	2.5	3.0	9.1	13.2
Germacrene B	0.5	0.4	0.6	0.5	5.1	2.7
Caryophyllene oxide	3.4	10.3	5.4	5.0	1.7	3.9

chemical (chromatographical) analysis or in specific cases by sensory evaluation (for instance, differentiation between mint linalool or carvone chemotype). To reduce this risk, one should consider careful raw material mixing to obtain the average batch of process input.

Before the consideration of technological features' influence on the quality of dried material, the issue of the material microbial safety has to be discussed. The optimal (guaranteeing the highest retention of expected phytochemicals or creating the expected sensory quality) for a particular raw material drying method will be worthless, if the protocol would not secure the product before microbial contamination. For instance, the traditional procedure, based on spreading the raw material on ground and performing shade or sun drying, brings high risks of microbial spoilage (Sagoo et al. 2009). The most common microbial contaminants of HSMPs include *Salmonella* ssp., *Bacillus cereus*, *Clostridium perfingers*, *Escherichia coli*, *Yersinia intermediata*, and/or *Shigella* ssp. (Sagoo et al. 2009; Sospedra et al. 2010). Thus, the preparation of plant material has to include the treatment dedicated for the reduction of potential contamination (like steam or dry heat treatment), and thereafter, the applied drying has to provide conditions, which would reduce the contamination risk. It is especially important when drying using microwaves when uneven temperature distribution can lead to occurrence of cold and hot spots that not only lead to quality decline but also can pose a microbiological risk, since in the cold spots, the temperature might be too low to eliminate harmful organisms (Vadivambal and Jayas 2010).

On the other hand, from the technological point of view, various parameters of the process may be adjusted and those possibilities have to be multiplied by the available drying methods. As it may be concluded from Figure 8.3, the technological factors having an influence on the drying procedure outcome are mutually dependent. To start with HSMPs phytochemicals sensitivity, volatility (e.g. EOs constituents), thermal robustness (e.g. constituents including heat-sensitive bindings), or oxidation robustness (e.g. lipids and vitamins) should be considered. The higher the content of fragile compounds in raw material the lower the quality of dried products. To manage this situation, one has to adjust the other factors, such as temperature, temperature distribution, drying time, and presence of oxygen. The last factor, oxygen presence, may be limited by performing procedures in vacuum or inert gas (e.g. nitrogen, CO_2) conditions.

Nevertheless, more common characteristics for almost all drying techniques are the time of the process and the temperature of the processed plant material (Öztekin and Martinov 2014; Rocha 2011). Those two factors are strongly connected, since higher the temperature of the process, the shorter the performance time (Łyczko et al. 2020). Taking two drying methods – convective drying (CD) and vacuum-microwave drying (VMD) – as examples, we may illustrate the influence of time and temperature on the quality of dried HSMPs. The studies performed on *Lavandula angustifolia* Mill. flowers (Łyczko et al. 2019a) and *Murraya koenigii* (Choo et al. 2020) leaves have shown that the increase of temperature while CD in 10°C intervals resulted in even twice the shortened time of process with each temperature level, although the quality of the dried products, measured, respectively, as total essential oil yield or total VOCs quantity was significantly affected. However, again the results

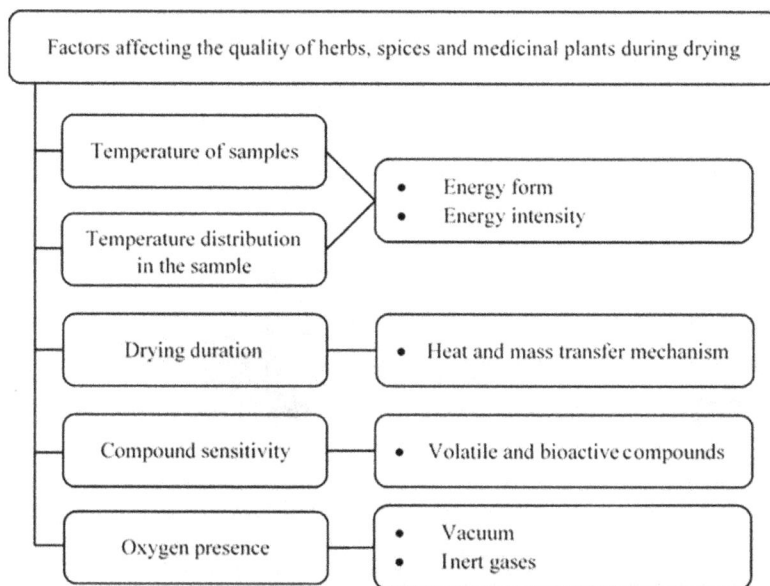

FIGURE 8.3   Factors affecting the quality of herbs, spices, and medicinal plants during drying.

are not correlated, as presented in Figure 8.4 – in both cases, the middle temperature level had presented the worst results. In the same research, the opposite trend was observed for VMD, when the middle drying parameters adjustment showed the best results in light of selected quality determinants. Similar issues were noticed in the case of *Cassia alata* (Chua et al. 2019b). Unfortunately, such a trend was not observed for each object. As an example, *Thymus vulgaris* L., which was also subjected to CD at three different temperatures and VMD with three magnetrons powers, the lowest adjustments of drying techniques gave the best VOCs retention (Calín-Sánchez et al. 2013). Therefore, it is clear that performing the preliminary laboratory experiments to establish the behavior of particular HSMPs genus, or even its variety, may be highly efficient.

However, as has been previously reported, temperature of the process might not be as important as actual temperature of the material during drying. Several factors can influence the temperature of the material, with the most important ones being energy form and intensity. Energy form depends on the heat transfer mode, which can include convection, conduction, radiation, and dielectric and combined mode (Acar et al. 2020). Depending on the energy form, certain issues might arise that can significantly influence the quality of dried material. For instance, during CD, the hot air flows through the material, heating it up from the outside, which can lead to shrinkage as well as to formation of the crest on the surface due to the high temperatures hindering water evaporation. On the other hand, during microwave drying, volumetric heating occurs which increases the temperature from the inside of the material. However, this method can be associated with uneven heating which results in the formation of cold and hot spots that can significantly decrease the quality of

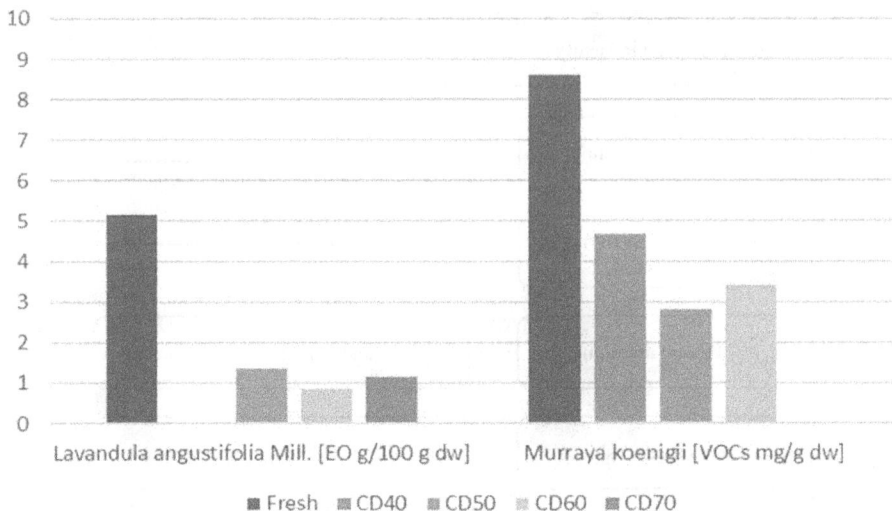

FIGURE 8.4 Changes in the essential oil yield or VOCs quantity in the chosen HSMPs caused by drying (Choo et al. 2020; Łyczko et al. 2019a).

the material. Another issue is energy intensity that needs to be carefully monitored by measurements of materials temperature in order to minimize degradation of thermolabile compounds present in HSMPs. Higher energy intensity of drying leads to shorter time of the process (Parlak 2015) but can also negatively affect the quality of the material.

Yet has it been defined, regarding the above consideration, which drying method will deliver the product with guaranteed quality? Yes, if the definition of guaranteed quality was determined as 'the highest, possible essential oil or VOCs retention'. Nevertheless, when the term quality has another meaning for a particular consumer or product user, the answer may be totally different. In the case of *L. angustifolia* Mill. flowers, as it was said, the optimal, in terms of essential oil yield was CD at 50°C, although this treatment was not the only one optimal for the sensory quality of dried product. For optimal sensory quality, also VMD at 360 W and combined method (consisting of CD predrying and VMD finish-drying).

8.3   NEEDS OF ADVANCED DRYING TECHNOLOGIES FOR HERBS, SPICES, AND MEDICINAL PLANTS

Several methods have been traditionally used for drying of HSMPs. Among the most common methods, among others, are sun and shade drying or hot air drying. Even though these methods are popular, they also have certain disadvantages, namely, exposure to harsh weather conditions, pests, and microbial spoilage along with long time which directly affects the quality of the dried material (Parlak 2015). Moreover, there is a high demand for good-quality products obtained in a sustainable and environmental-friendly manner. Therefore, there is a significant need for new and advanced drying technologies that would satisfy these needs.

Advances in drying technologies for HSMPs should be especially aimed at achieving the highest quality of materials after the process. As already mentioned in the previous section, acceptable quality of HSMPs can be achieved by minimizing the negative effects of temperature, light, as well as environment (presence of oxygen) and time of drying. However, new technologies also need to be cost-effective and reduce the environmental impact. It can be maintained by applying renewable energy technologies in the process as well as using hybrid drying systems which combine the advantages of different drying methods in order to achieve the synergistic effect in terms of quality as well as time and consequently energy efficiency of the process.

HSMPs are especially vulnerable to high temperatures. Therefore, emerging and advanced drying methods for these types of products should operate in moderate conditions that will ensure the highest quality of materials and minimize losses of thermolabile substances. One of the ways can be application of reduced pressure or vacuum that will decrease the temperature of water evaporation and also increase the internal pressure gradient which will improve the water diffusion and accelerate the drying rate (Hirun et al. 2014). Another factor that needs to be considered is oxygen presence. As already discussed, oxidation of the material should be avoided to improve the quality; thus methods that use, other than hot air, mediums such as superheated steam (Patel and Bade 2021; Romdhana et al. 2015) or supercritical CO_2 (Michelino et al. 2018; Zambon et al. 2018) should be taken into account when discussing emerging drying methods for HSMPs. Moreover, the presence of oxygen can be minimized by operating under vacuum, however, reduced pressure during drying might be connected with high loses of volatile compounds. Therefore, methods that allow capturing of these compounds should be considered.

However, what is most important is monitoring the parameters of the materials during drying and not only processing conditions. For instance, application of microwaves can lead to local overheating due to the uneven heat distribution in the sample. One of the ways of overcoming this problem is application of intermittent processes that will enable moisture and heat distribution during intervals and help maintain similar temperature in the whole volume of the material in the course of drying (Duc Pham et al. 2019).

Another issue that needs to be addressed is drying kinetics. During drying, two periods can be distinguished, constant rate period and falling rate period. In the constant rate period, more intense energy can be supplied to the material as the cooling effect provided by the intense evaporation will control and maintain moderate temperature of the material. Then, in the falling rate period, when the easily accessible water is already removed, application of high-intensity heating might lead to significant degradation in quality (Orphanides et al. 2016). On the other hand, too mild drying parameters might unreasonably prolong the process leading to significant increase in energy consumption. Therefore, for new drying technologies, it is necessary to consider methods that allow adjustment of the processing conditions according to the stage of drying. Another way of controlling the temperature and quality of the material could be application of combined drying where a certain sequence of methods is applied in a way that will improve the drying rate by switching the heat transfer mode, operation times, and technologies. However, these methods require extensive research in order to precisely determine the moment when methods should be changed or adjusted to obtain the best quality dried material (Calín-Sánchez et al. 2014). Another challenge associated with drying is high energy consumption. Therefore, new and advanced drying methods need not only to provide high-quality products but also minimize energy consumption and reduce the environmental impact of drying. In the review by Acar, Dincer, and Mujumdar (2020), several new and advanced drying technologies have been discussed in detail in terms of thermal efficiency and energy efficiency, operating costs, as well as primal heat and electricity use. Based on these factors, sustainability ranking was presented that showed which methods are promising in terms of clean, affordable, reliable, and efficient drying. Minimizing the environmental impact of industrial processes has become a pressing matter, and thus, methods that allow regeneration of the drying medium or increase the use of renewable energies should be prioritized in development of emerging and advanced drying methods.

8.4 EMERGING AND ADVANCED DRYING TECHNOLOGIES FOR COST-EFFECTIVE AND ENHANCED QUALITY OF DRIED PRODUCTS

Drying should be performed in a way that will minimize the quality losses during the process. It can be maintained by managing the drying methods and parameters but also by applying different treatments such as osmotic dehydration, ultrasounds, pulsed electric field, and others that will positively influence the quality of the material while reducing the moisture content or increasing the microbial safety. Therefore, these treatments are discussed in detail in the following sections.

8.4.1 TREATMENT METHODS

Osmotic dehydration (OD) is a process driven by the difference in osmotic pressure between the material and hypertonic osmotic solution. As a result, a mass transfer occurs which leads to water reduction, but at the same time, substances from the solution are transported to the material increasing solid gain during the process. OD is characterized by limited cell damage, high retention of nutrients, and low energy requirements compared to conventional drying (Champawat et al., n.d.). However, this treatment needs to be coupled with another drying method to obtain stable and safe material. Eucalyptus degulpta was dried using ultrasound-assisted OD and hot air drying (Sukumaran et al. 2021) (Table 8.3). In another study, it was found that application of ultrasound-assisted OD as a pretreatment led to two times higher antioxidant activity as well as total phenolic content compared to CD (Chua et al. 2021). OD is a method that can be used in order to retain the taste and nutritional value (Alam et al. 2018). Also, OD was found to increase the amount of undesirable aromatic compounds such as myristicin in nutmeg (Rahman et al. 2018). OD can be used together with ultrasounds and vacuum such as in the studies on garlic slices where it led to high quality in terms of i.a. allicin content (Feng et al. 2019). Furthermore, the process can be conducted at high pressure (Dash et al. 2019). Osmosonication treatment of ginger slices resulted in an increase in antioxidant activity as well as bioactive and phytochemical compounds (Osae et al. 2019). OD was also applied in the case of green chili (Haque et al. 2019). As can be seen, this method can significantly affect the quality of the material along with reduction of moisture content.

Another type of pretreatment is **blanching** which due to the high temperatures leads to killing of pathogenic microorganisms, browning enzymes inactivation, as well as microstructural changes inside the material (Majumder et al. 2021). Boiling temperature, type of heating (conventional, ohmic heating), medium (steam, water), or surface area affect the blanching process. Blanching inactivates enzymes that would normally result in color, flavor, and texture deterioration. Moreover, application of blanching as a pretreatment was reported to minimize the amount of toxic myristicin in nutmeg (Putsakum et al. 2020). Blanching can be applied along other treatments such as vacuum, where the material is placed in a vacuum bag that is then put into hot water, or application of chemical agents, i.e. sodium metabisulfite and citric acid or potassium metabisulfite (Amoah et al. 2020), can be added during the process which decreases the microbial load. Also, the heat necessary to perform the process can be delivered by water, steam, or microwaves. Water blanching is the most popular method used in the case of several herbs such as *Anacardium occidentale* and *Piper betle* (Tan and Chan 2014), pepper (Weil et al. 2017), or garlic (Zhang et al. 2021a). However, this method is associated with losses of water-soluble compounds (Gan et al. 2017). Steam blanching, on the other hand, was previously used in the case of marjoram and parsley (Dadan et al. 2018; Kaiser et al. 2013). Another modification of blanching is high-humidity hot air impingement blanching (HHAIB). This method combines the advantages of steam blanching and hot air in order to obtain uniform heating in the material without the loss of water-soluble compounds.

Pulsed electric field (PEF) uses external electric field that when applied to plants results in poration of cell membranes which can be permanent or reversible. Permeabilization significantly reduces drying time as a result of increase in moisture coefficient (Thamkaew et al. 2021). Consequently, application of PEF before drying reduces the time of thermal treatment which is crucial when handling thermosensitive products such as HSMPs. Moreover, PEF leads to a higher retention of carotenoid pigments after drying such as in the studies on red pepper (Won et al. 2015).

Similarly, to PEF treatment, **high-pressure processing (HPP)** also leads to permeabilization of cell membrane which affects moisture diffusivity in the material and as a result shortens the time of drying. Also, due to high pressure, softening of the tissues occurs, consequently changing the cellular matrix and permeabilization of cell membrane. HPP was previously applied before drying of ginger (George et al. 2018). Such pretreatment significantly increased the drying rate which resulted in shortening of drying time. Also, HPP increased oleoresin yield in the material. This method is usually used as a sterilization method of heat-sensitive materials (Yang et al. 2018). Moreover, application of high-pressure pretreatment can improve antioxidant activity (Vega-Gálvez et al. 2011).

Ultrasounds are short-length acoustic waves that can be used as a pretreatment method before drying or as an assisting treatment during drying. Application of ultrasounds leads to rapid compressions and expansions of the material which is called sponge effect. As a result, mass transfer accelerates with only little temperature increase. Moreover, formation of micropores occurs which minimizes hardening

TABLE 8.3

Quality of Materials Treated by Combined Drying

Material	Drying Method	Quality Compared to Convective Drying	References
Garlic	CD 60°C CPD-VMFD (CD 60°C+VMD 240, 360 and 480 W)	Total phenolic content not improved Higher antioxidant activity	Calín-Sánchez et al. (2014)
Thai basil leaves	CD VMD CPD-VMFD	Best quality: fresh, intense aroma → combined drying	Łyczko et al. (2020)
Coriander leaves	CD VMD CPD-VMFD	Highest quality for combined method	Łyczko et al. (2021)
Hemp flowers	CD VMD CPD-VMFD	Lower losses of volatile compounds for VMD Better sensory characteristic for VMD	Kwaśnica et al. (2020)
True lavender leaves	CD (50°C, 60°C, 70°C) VMD (240, 360, 480 W) CPD-VMFD (CD 60°C, VMFD 480 W)	Highest recovery of essential oils for VMD Highest reduction in total essential oil and major volatile constituents for CPD-VMFD	Łyczko et al. (2019b)
Thyme	CD (40°C, 50°C, 60°C) VMD (240, 360, 480 W) CPD-VMFD (40°C, 50°C, 60°C and 240, 360, 480 W)	Highest concentration of volatile compounds for VMD 240 and 360 W CPD-VMFD high sensory quality High intensity of freshness for VMD and CPD-VMFD	Calín-Sánchez et al. (2013)
Sweet basil	CD (40°C, 50°C, 60°C) VMD (240, 360, 480 W) CPD-VMFD (40°C, 50°C, 60°C and 240, 360, 480 W)	CPD-VMFD higher quantity of volatile compounds and high quality in sensory profile	Calín-Sánchez et al. (2012)
Cassia alata	CD (40°C, 50°C, 60°C) VMD (6, 9, 12 W/g) CPD-VMFD (CD 50°C for 90 min followed by VMFD at 9 W/g)	Lower antioxidant activity, total phenolic content, and total volatile concentration Higher phytosterol concentration than CD40	Chua et al. (2019b)

CD, convective drying; VMD, vacuum-microwave drying; CPD-VMFD, convective predrying followed by vacuum-microwave finishing drying.

of the surface of the material during drying, consequently limiting disturbances in water removal (Nowacka and Wedzik 2016). This method can be used along with other pretreatments, such as already discussed OD, or with drying methods, namely, CD (Schössler et al. 2012), supercritical CO_2 drying (Michelino et al. 2018), and others. Research revealed that application of high-intensity ultrasounds during drying can result in increased mass transfer; however, it also depends on the drying parameters (air velocity and drying temperature), suggesting the need for further studies regarding this issue (Rodríguez et al. 2014).

8.4.2 Advanced Drying Methods

Aside from the application of different treatments that can influence the quality and energy consumption of drying, there are several advances in existing drying methods that either reduce the energy consumption and cost of drying or significantly improve the quality of dried materials. Reduction of moisture content is the main aim of drying and is believed to ensure microbial safety of dried materials. However, even though microbial growth is hindered or even prevented in dried products, pathogenic spores and vegetative cells are still present in the material and can pose a significant risk after rehydration. While additional antimicrobial treatment is not necessary in the case of fruits and vegetables, it is often practiced in the case of herbs and spices. Therefore, several nonthermal methods can be applied such as irradiation, radio frequency heating, and ohmic heating (Qiu et al. 2020).

One of the methods that combine the antimicrobial effect with reduction of water activity is application of **supercritical carbon dioxide drying** (scCO$_2$). Supercritical CO$_2$ drying is known as a nonthermal method which ensures microbial inactivation in the dried material and was previously applied in drying of coriander leaves (Zambon et al. 2018) and basil (Bušić et al. 2014). The process consists of three stages, namely, pressurization, drying, and depressurization. The method uses CO$_2$ which is an inert gas, supercritical at low temperature (31.1°C), nontoxic and nonflammable, as well as easily available and inexpensive. Furthermore, CO$_2$ can be easily recirculated and reused which limits the costs and environmental impact. Moreover, drying using supercritical fluids, due to the lack of vapor–liquid interfaces, reduces the capillary stress and consequently leads to less shrinkage and higher preservation of material structure (Khalloufi et al. 2010). Supercritical carbon dioxide drying can also be used together with other techniques such as high-power ultrasounds in order to increase the efficiency of drying as well as improve microbial inactivation (Michelino et al. 2018). Application of scCO$_2$ drying ensures the high quality of dried product; however, this method has not been widely used in drying of HSMPs, and there is a need for further studies considering the quality in terms of, among others, volatile compounds and phenolic compounds retention after drying using this method.

Superheated steam drying (SSD) uses superheated steam as a drying medium and is performed in an insulated drying chamber equipped with a steam trap that reduces the condensation in the chamber. Elimination of oxygen by using superheated steam as a drying medium results in high quality of dried materials in terms of color, porosity, rehydration behavior, and higher nutrient preservation. Furthermore, removal of air in the process inhibits enzymatic browning; however, high temperatures can still lead to caramelization or Maillard reactions which impact the color of dried material (Romdhana et al. 2015). Application of superheated steam as a drying medium is also advantageous in terms of energy consumption and environmental impact as the latent heat of evaporation can be reused by condensation of exhaust steam (Sehrawat et al. 2016). SSD has lower energy consumption than hot air drying; however, this method is mostly used for materials with low-to-moderate moisture content. High energy efficiency when drying using this method is mostly due to the recirculation of exhaust steam or heat recovery (Patel and Bade 2021). However, according to the study by Acar, Dincer, and Mujumdar (2020), SSD scored poorly in sustainability ranking due to the high operating costs and high final heat use, among other things. One of the main drawbacks when using this method for drying of thermolabile materials is high temperature of the process (>100°C). Application of low pressure during the process can compensate for this difficulty as it lowers the temperature of evaporation. **Low-pressure superheated steam drying (LPSSD)** method was previously used, among others, in the case of onion (Sehrawat and Nema 2018). Onion treated by **LPSSD** showed better color, rehydration characteristics, as well as retained high pungency compared to vacuum and hot air drying.

One of the electromagnetic techniques used in HSMP drying is **microwave drying (MD)**. During MD, volumetric heating occurs in the material due to oscillation of water molecules as a result of electromagnetic waves penetrating the material. This leads to increased drying rate and consequently shorter time of drying, increased energy efficiency, and reduced operational costs compared to hot air drying. However, this method also generates some challenges caused by uneven heating (especially local overheating) and disrupted mass transfer which directly affects the quality of dried materials (Vadivambal and Jayas 2010). One of the advancements in MD is **pulsed microwave drying** that due to the intervals between heating periods enables more even distribution of heat during MD, consequently leading to higher quality of dried materials (Duc Pham et al. 2019). What is also incredibly interesting is application of other techniques to improve the quality of dried products, such as carbonic maceration which was previously used in the microwave drying of chili and was found to increase the drying rate, antioxidant activity, and total phenols and vitamin c content compared to MD without pretreatment (Liu et al. 2014).

Infrared drying (IR) also works on the basis of electromagnetic energy from the infrared wavelengths that get absorbed by the moisture content in the material as a result generating heat and therefore leading to water evaporation. Among the advantages of IR can be listed: small size of the equipment, significant energy savings, and the ability to assemble with other drying equipment such as convective or MD (Qiu et al. 2020). Furthermore, infrared drying exhibits a higher efficiency and drying rate compared to hot air drying (Miraei Ashtiani et al. 2017). IR was reported to lead to higher retention of gingerols compared to other methods (An et al. 2016). Furthermore, this method is suitable for obtaining high quality materials with relatively low operating costs and high efficiency of drying.

Refractance window drying (RW) is one of the most promising drying methods since it operates in mild conditions, i.e. temperature around 30°C and atmospheric pressure, which helps to limit the degradation of thermolabile compounds during drying (Nindo and Tang 2007). The method uses three heat transfer modes, such as convection, conduction, and radiation, which contribute to short drying time makes this method a low-cost alternative to other drying methods (Calín-Sánchez et al. 2020). Moreover, RW ensures high quality of dried materials. In the studies on aloe vera, it was found that the samples dried using RW presented similar color to freeze-dried materials, even after storage (Nindo and Tang 2007).

Another electromagnetic method used in drying of HSMP is **radio frequency drying (RF)**. The method uses a radio frequency generator to create alternating electric fields. As a result, polar molecules present in the food products rotate rapidly creating heat that leads to water evaporation. This method has been frequently used along with CD (Calín-Sánchez et al. 2020). However, in the study on drying of dill, it was found that RF drying resulted in longer drying time and lower temperature of the material during drying compared to hot air drying. However, retention of ascorbic acid and chlorophyll content was lower than in the case of low humidity air drying (Madhava Naidu et al. 2016).

8.4.3 HYBRID DRYING

Hybrid drying can be adopted to combine the advantages of several single drying methods resulting in shorter time of the process and improved quality. This synergistic effect can be achieved by either simultaneous application of two or more drying mechanisms at once or by using **combined drying methods** when appropriate sequence of different drying methods is applied. Some examples of hybrid drying methods are discussed further in this section.

Hybrid drying consisting of **convective and microwave drying** has been previously studied by Łechtańska, Szadzińska, and Kowalski (2015) and showed high vitamin C retention, shortened drying time, and lower energy consumption compared to CD. Microwave drying can also be conducted under vacuum. **Microwaves-vacuum drying** (MVD) combines the advantages of volumetric heating occurring when microwaves are applied and fast mass transfer caused by pressure gradient when the process is performed under vacuum (Hirun et al. 2014).

The application of **intermittent microwave convective drying (IMCD)** can be especially important as it overcomes the main disadvantages of both methods, i.e. reduces the time of traditional CD and at the same time due to intermittent heating eliminates the challenges associated with uneven heat distribution during MD. Periodic application of microwaves limits the occurrence of local overheating as during the intervals the moisture and heat transfer occur, which leads to uniform heating of the material in the whole volume. As a result, not only higher quality materials can be obtained but also energy consumption can be reduced as the time of the process is significantly decreased. Overall, **intermittent drying** operates with variations in supply of thermal energy which can be achieved by changes in air temperature, humidity, and pressure or even energy input in the course of drying (Kumar et al. 2014). As a result, materials dried using this method exhibit higher quality, especially in terms of color since browning is reduced. Moreover, due to the intervals, temperature and heat distribution are more even in the material which results in decreased case hardening and reduced energy consumption (Duc Pham et al. 2019).

Another hybrid drying method is **infrared convective drying** which combines the advantages of hot air drying and infrared radiation. This method was previously found to improve the quality of food products along with the reduction of energy consumption during drying (Łechtańska et al. 2015). This method was also used in the study on physical properties of nutmeg (Ps et al. 2018). **Microwave infrared drying (MWIR)**, on the other hand, was found to reduce the energy consumption in the studies on red bell pepper (Kowalski and Mierzwa 2011). **Far-infrared microwave vacuum drying (FIRMVD)** was previously applied in the drying of red chili (Saengrayap et al. 2015). The paper studied the quality changes of the material depending on the microwave and far-infrared power as well as pressure during drying. The study showed that careful consideration of these parameters is especially important when two heat modes are simultaneously applied. Among hybrid drying methods using electromagnetic techniques, aside from microwaves, is also **radio frequency drying under vacuum** or **radio frequency convective tray drying (RFCTD)** where additional heat source improves water evaporation and as a result water diffusivity, even at low temperatures (Madhava Naidu et al. 2016).

Advances in drying can include application of additional equipment in order to minimize the environmental impact of drying as well as reduce energy consumption and other emissions. **Solar- and heat-pump-assisted drying** are among the most popular methods aimed at reducing the costs and environmental impacts of drying. According to Jin et al. (2018), solar energy can be utilized in drying processes through (1) direct or open air drying, (2) indirect or convective solar drying, or (3) hybrid drying. Direct or open air solar drying comes with several challenges such as low drying rate, poor quality, microbial contamination, high labor costs, and product losses due to the presence of insects and birds (Qiu et al. 2020). Indirect solar drying limits the described challenges but still requires a long time for drying. Hybrid drying systems, on the other hand, use solar energy as a source of energy to power other equipment that is used for drying minimizing energy costs. Interesting solution for reducing the energy consumption during drying is utilization of waste heat by implementation of heat pumps. **Heat pumps** during drying using various methods can reduce the negative impact of drying and minimize energy consumption (Calín-Sánchez et al. 2020). Heat pumps can be integrated with solar dryers, fluidized bed dryers, as well as far-infrared radiation drying. Heat pumps integrated with solar dryer were used in the drying of saffron and resulted in a reduction of drying time and consequently energy consumption together with improvement of electric efficiency (Phoungchandang et al. 2009). **Far-infrared radiation-assisted heat pump (HP-FIR)** was previously applied in the drying of garlic where this method reduced the time of the process compared to traditional heat pump or far-infrared drying (Younis et al. 2018). Moreover, study by Acar et al. showed that heat pumps have high energy efficiency, only lower than adsorption-mediated drying (Acar et al. 2020).

8.4.4 Combined Drying Methods

Combined drying consists of two or more drying methods applied one after the other in order to reduce the time of the process and improve the quality of dried materials. The main objective is to apply a sequence of different drying methods in a way that will lead to the synergistic effect. One of the combined methods is convective predrying followed by vacuum-microwave finishing drying which is presented in Table 8.3.

Convective predrying is often applied in the first stage of drying as during this method high quantities of free water can be easily evaporated from the surface of the material through external diffusion. Then, VMD is applied in the second stage of drying to increase the drying rate (compared to CD), shorten duration of the process, as well as improve the quality of dried material. The quality of HSMPs dried using this method is achieved by reduced time of oxygen exposure as well as improved evaporation due to the volumetric heating and reduced pressure during the second stage of drying. Moreover, application of VMD in the falling rate period can significantly reduce the energy consumption compared to CD and increase energy savings as shown in the studies on garlic (Calín-Sánchez et al. 2014). Furthermore, application of combined CD-VMD leads to obtaining fresh and intense aroma of herbs such as Thai basil (Łyczko et al. 2020). Similarly, higher quality of dried coriander leaves can be achieved by combined methods compared to CD and VMD (Łyczko et al. 2021). On the other hand, combined drying of hemp leaves resulted in more pronounced degradation of chemical composition underlining the importance of considering the individual properties of plant raw materials when determining the best drying methods (Kwaśnica et al. 2020). Furthermore, combined CD-VMD drying leads to higher energy saving and lower energy consumption compared to CD; however, energy consumption is still lower than for single VMD (Calín-Sánchez et al. 2014; Chua et al. 2019b; Stępień et al. 2019).

Another combined method can be **intermittent superheated steam drying coupled with low air drying**. During this method, SSD is applied at the beginning of drying to ensure high evaporation rate and then, before degradation of the material occurs, low air drying is applied. This method not only provides better quality material but also can use the exhaust steam to heat the drying air in the second stage of drying, reducing the energy consumption of the process. Another example of combined drying is **microwave and far-infrared drying**. This method was previously used in the drying of ginseng. The study showed that this combination resulted in faster drying rate, lower shrinkage, and lower color difference when compared to single far-infrared drying. Moreover, the study evaluated the optimal switching point between the methods (Ning et al. 2019), which shows that the moment of change during combined methods needs to be carefully considered.

8.5   FUTURE DEVELOPMENTS

The present emerging and new drying technologies offer some solutions to the identified challenges associated with drying of HSMPs; however, there is still a lot that needs to be done to improve the current situation and develop drying methods that will respond to both producers and consumers' demands and needs. First and foremost, investments in the R&D of drying technologies should be continued and intensified, especially since drying is one of the most important and, at the same time, most energy-consuming process (Mujumdar 2007). Therefore, methods combining the utilization of renewable energies and recirculation of the drying medium should be prioritized. Furthermore, some promising technologies studied in the laboratory are not scalable or require high investment costs in industry conditions; thus intense research to find new and more adaptable methods and solutions is essential.

Future developments that should be considered can be listed as follows:

- Quality of HSMPs is of the highest priority; therefore, the properties of the material during drying should be carefully controlled. To achieve that, collecting data and monitoring process parameters and properties of the material in the course of drying, especially online, should be incorporated in the designing and development of new methods. Moreover, advanced methods aimed at evaluation of physical and chemical properties of the material in the course of drying should be considered. New technologies such as low-field NMR technology as well as electronic nose would be beneficial in terms of ensuring high quality of dried materials (Yang et al. 2018).
- Hybrid drying methods and systems seem to be a most promising way of achieving high-quality products with reduced energy and environmental costs. In order to ensure correct parameters and switching times, especially in the case of combined drying, internal moisture content and distribution along with water mobility could be monitored in real life using, for example, hyperspectral technology (Liu et al. 2017). Moreover, modeling of the drying process should include application of neural networks and artificial intelligence to better understand and predict the behavior of the material under various conditions.
- Additional research on the changes of the structure, color, and overall quality of the HSMPs in the course of drying should be performed as there is insufficient amount of research conducted with these materials, especially in terms of quality changes occurring in the course of drying. Therefore, integration of different types of sensors into the drying installations allowing constant monitoring of the quality of the material as well as development of easily applicable quality markers should be considered in the future.

To sum up, the most important is finding a balance between the time, energy consumption, and cost of the process by maintaining high quality of dried materials. Application of abovementioned developments should result in achieving this goal.

8.6 CONCLUDING REMARKS

HSMPs are materials rich in valuable compounds that have the ability to limit or counteract pathogens in human body. Therefore, maintaining the quality of these materials obtained in the drying process is imperative. Furthermore, it is essential for the proposed drying technologies to incorporate innovative solutions aimed at achieving the best quality of the final product. Dried HSMPs should belong to the premium group of products in which the economic aspect plays a minor role. Development of technologies based on several drying methods and the possibility of online control of the properties of dried products might be costly, but, at the end, it will bring invaluable benefits to the health of the society.

REFERENCES

Acar, Canan, Ibrahim Dincer, and Arun Mujumdar. 2020. "A Comprehensive Review of Recent Advances in Renewable-Based Drying Technologies for a Sustainable Future." *Drying Technology*: 1–27. https://doi.org/10.1080/07373937.2020.1848858.

Alam, Md Sahin, M. Kamruzzaman, Sultana Anjuman Ara Khanom, Mohammad Robel Hossen Patowary, Md Toufiq Elahi, Md Hasanuzzaman, and Dipak Kumar Paul. 2018. "Quality Evaluation of Ginger Candy Prepared by Osmotic Dehydration Techniques." *Food and Nutrition Sciences* 9 (4): 376–89. https://doi.org/10.4236/fns.2018.94030.

Amoah, Roseline Esi, Faustina Dufie Wireko-Manu, Ibok Oduro, Firibu Kwesi Saalia, and William Otoo Ellis. 2020. "Effect of Pretreatment on Physicochemical, Microbiological, and Aflatoxin Quality of Solar Sliced Dried Ginger (Zingiber Officinale Roscoe) Rhizome." *Food Science & Nutrition* 8 (11): 5934–42. https://doi.org/10.1002/fsn3.1878.

An, Kejing, Daobang Tang, Jijun Wu, Manqin Fu, Jing Wen, Gengsheng Xiao, and Yujuan Xu. 2019. "Comparison of Pulsed Vacuum and Ultrasound Osmotic Dehydration on Drying of Chinese Ginger (*Zingiber officinale* Roscoe): Drying Characteristics, Antioxidant Capacity, and Volatile Profiles." *Food Science & Nutrition* 7 (8): 2537–45. https://doi.org/10.1002/fsn3.1103.

An, Kejing, Dandan Zhao, Zhengfu Wang, Jijun Wu, Yujuan Xu, and Gengsheng Xiao. 2016. "Comparison of Different Drying Methods on Chinese Ginger (Zingiber officinale Roscoe): Changes in Volatiles, Chemical Profile, Antioxidant Properties, and Microstructure." *Food Chemistry* 197 (Pt B) (April): 1292–300. https://doi.org/10.1016/j.foodchem.2015.11.033.

Bušić, Arijana, Aleksandra Vojvodić, Draženka Komes, Cynthia Akkermans, Ana Belščak-Cvitanović, Maarten Stolk, and Gerard Hofland. 2014. "Comparative Evaluation of CO_2 Drying as an Alternative Drying Technique of Basil (Ocimum Basilicum L.) — The Effect on Bioactive and Sensory Properties." *Food Research International* 64 (October): 34–42. https://doi.org/10.1016/j.foodres.2014.06.013.

Calín-Sánchez, Ángel, Adam Figiel, Krzysztof Lech, Antoni Szumny, and Ángel A. Carbonell-Barrachina. 2013. "Effects of Drying Methods on the Composition of Thyme (Thymus Vulgaris L.) Essential Oil." *Drying Technology* 31 (2): 224–35. https://doi.org/10.1080/07373937.2012.725686.

Calín-Sánchez, Ángel, Adam Figiel, Aneta Wojdyło, Marian Szarycz, and Ángel A. Carbonell-Barrachina. 2014. "Drying of Garlic Slices Using Convective Pre-Drying and Vacuum-Microwave Finishing Drying: Kinetics, Energy Consumption, and Quality Studies." *Food and Bioprocess Technology* 7 (2): 398–408. https://doi.org/10.1007/s11947-013-1062-3.

Calín-Sánchez, Ángel, Krzysztof Lech, Antoni Szumny, Adam Figiel, and Ángel A. Carbonell-Barrachina. 2012. "Volatile Composition of Sweet Basil Essential Oil (Ocimum basilicum L.) as Affected by Drying Method." *Food Research International* 48 (1): 217–25. https://doi.org/10.1016/j.foodres.2012.03.015.

Calín-Sánchez, Ángel, Leontina Lipan, Marina Cano-Lamadrid, Abdolreza Kharaghani, Klaudia Masztalerz, Ángel A. Carbonell-Barrachina, and Adam Figiel. 2020. "Comparison of Traditional and Novel Drying Techniques and Its Effect on Quality of Fruits, Vegetables and Aromatic Herbs." *Foods* 9: 1–27.

Champawat, P. S., Waghmode, A. S., Mudgal, V. D., & Madhu, B. (2019). Effect of temperature and salt concentration during osmotic dehydration of garlic cloves. *International Journal of Seed Spices*, 9, 61–66.

Chen, Heng-Wen, Ben-Jun Wei, Xuan-Hui He, Yan Liu, and Jie Wang. 2015. "Chemical Components and Cardiovascular Activities of Valeriana spp." *Evidence-Based Complementary and Alternative Medicine* 2015 (December): 947619. https://doi.org/10.1155/2015/947619.

Choo, Choong Oon, Bee Lin Chua, Adam Figiel, Klaudiusz Jałoszyński, Aneta Wojdyło, Antoni Szumny, Jacek Łyczko, and Chien Hwa Chong. 2020. "Hybrid Drying of Murraya Koenigii Leaves: Energy Consumption, Antioxidant Capacity, Profiling of Volatile Compounds and Quality Studies." *Processes* 8 (2): 240. https://doi.org/10.3390/pr8020240.

Chua, Lisa Y. W., Chien Hwa Chong, Bee Lin Chua, and Adam Figiel. 2019a. "Influence of Drying Methods on the Antibacterial, Antioxidant and Essential Oil Volatile Composition of Herbs: A Review." *Food and Bioprocess Technology* 12 (3): 450–76. https://doi.org/10.1007/s11947-018-2227-x.

Chua, Lisa Yen Wen, Bee Lin Chua, Adam Figiel, Chien Hwa Chong, Aneta Wojdyło, Antoni Szumny, and Krzysztof Lech. 2019b. "Characterisation of the Convective Hot-Air Drying and Vacuum Microwave Drying of Cassia Alata: Antioxidant Activity, Essential Oil Volatile Composition and Quality Studies." *Molecules* 24 (8): 1625. https://doi.org/10.3390/molecules24081625.

Chua, B. L., Y. C. Khor, A. Ali, and H. Ravikumar. 2021. "Influence of Ultrasound-Assisted Osmotic Dehydration Pre-Treatment on Total Phenolic Content, Antioxidant Capacity and P-Cymene Content of Eucalyptus Deglupta." *Journal of Tropical Forest Science* 33 (2): 149–59. https://www.jstor.org/stable/27007563.

Dadan, Magdalena, Katarzyna Rybak, Artur Wiktor, Malgorzata Nowacka, Joanna Zubernik, and Dorota Witrowa-Rajchert. 2018. "Selected Chemical Composition Changes in Microwave-Convective Dried Parsley Leaves Affected by Ultrasound and Steaming Pre-Treatments – An Optimization Approach." *Food Chemistry* 239 (January): 242–51. https://doi.org/10.1016/j.foodchem.2017.06.061.

Dash, Kshirod K., V. M. Balasubramaniam, and Shreya Kamat. 2019. "High Pressure Assisted Osmotic Dehydrated Ginger Slices." *Journal of Food Engineering* 247 (April): 19–29. https://doi.org/10.1016/j.jfoodeng.2018.11.024.

Duc Pham, Nghia, Md Imran H. Khan, M. U. H. Joardder, M. M. Rahman, Md. Mahiuddin, A. M. Nishani Abesinghe, and M. A. Karim. 2019. "Quality of Plant-Based Food Materials and Its Prediction during Intermittent Drying." *Critical Reviews in Food Science and Nutrition* 59 (8): 1197–211. https://doi.org/10.1080/10408398.2017.1399103.

El-Saber Batiha, Gaber, Amany Magdy Beshbishy, Lamiaa G. Wasef, Yaser H. A. Elewa, Ahmed A. Al-Sagan, Mohamed E. Abd El-Hack, Ayman E. Taha, Yasmina M. Abd-Elhakim, and Hari Prasad Devkota. 2020. "Chemical Constituents and Pharmacological Activities of Garlic (Allium sativum L.): A Review." *Nutrients* 12 (3): E872. https://doi.org/10.3390/nu12030872.

Feng, Yabin, Xiaojie Yu, Abu ElGasim A. Yagoub, Baoguo Xu, Bengang Wu, Lei Zhang, and Cunshan Zhou. 2019. "Vacuum Pretreatment Coupled to Ultrasound Assisted Osmotic Dehydration as a Novel Method for Garlic Slices Dehydration." *Ultrasonics Sonochemistry* 50 (January): 363–72. https://doi.org/10.1016/j.ultsonch.2018.09.038.

Gan, Haozhe, Erin Charters, Robert Driscoll, and George Srzednicki. 2017. "Effects of Drying and Blanching on the Retention of Bioactive Compounds in Ginger and Turmeric." *Horticulturae* 3 (1): 13. https://doi.org/10.3390/horticulturae3010013.

George, Jincy M., Halagur B. Sowbhagya, and Navin K. Rastogi. 2018. "Effect of High Pressure Pretreatment on Drying Kinetics and Oleoresin Extraction from Ginger." *Drying Technology* 36 (9): 1107–16. https://doi.org/10.1080/07373937.2017.1382505.

Glas, Joris J., Bernardus C. J. Schimmel, Juan M. Alba, Rocío Escobar-Bravo, Robert C. Schuurink, and Merijn R. Kant. 2012. "Plant Glandular Trichomes as Targets for Breeding Or Engineering of Resistance to Herbivores." *International Journal of Molecular Sciences* 13 (12): 17077–103. https://doi.org/10.3390/ijms131217077.

Haque, M. R., M. M. Hosain, M. S. Awal, and M. M. Kamal. 2019. "Optimization of Process Variables Affecting Osmotic Dehydration of Green Chili in Sucrose Solution by Response Surface Methodology." *American Journal of Food Science and Technology* 7 (3): 79–85. https://doi.org/10.12691/ajfst-7-3-2.

Hirun, Sathira, Niramon Utama-Ang, and Paul D. Roach. 2014. "Turmeric (Curcuma Longa L.) Drying: An Optimization Approach Using Microwave-Vacuum Drying." *Journal of Food Science and Technology* 51 (9): 2127–33. https://doi.org/10.1007/s13197-012-0709-9.

Husnu Can Baser, K., and Gerhard Buchbauer. 2016. *Handbook of Essential Oils: Science, Technology, and Applications*. 3rd ed. CRC Press. https://www.routledge.com/Handbook-of-Essential-Oils-Science-Technology-and-Applications/Baser-Buchbauer/p/book/9780815370963.

Jin, Wei, Arun S. Mujumdar, Min Zhang, and Weifeng Shi. 2018. "Novel Drying Techniques for Spices and Herbs: A Review." *Food Engineering Reviews* 10 (1): 34–45. https://doi.org/10.1007/s12393-017-9165-7.

Jin, Wei, Min Zhang, Weifeng Shi, and Arun S. Mujumdar. 2020. "Recent Developments in High-Quality Drying of Herbs and Spices." In *Herbs, Spices and Medicinal Plants*, edited by Mohammad B. Hossain, Nigel P. Brunton, and Dilip K. Rai, 1st ed., 45–68. Wiley. https://doi.org/10.1002/9781119036685.ch3.

Kaiser, Andrea, Reinhold Carle, and Dietmar R. Kammerer. 2013. "Effects of Blanching on Polyphenol Stability of Innovative Paste-like Parsley (Petroselinum Crispum (Mill.) Nym Ex A. W. Hill) and Marjoram (Origanum Majorana L.) Products." *Food Chemistry* 138 (2): 1648–56. https://doi.org/10.1016/j.foodchem.2012.11.063.

Khalloufi, S., C. Almeida-Rivera, and P. Bongers. 2010. "Supercritical-CO_2 Drying of Foodstuffs in Packed Beds: Experimental Validation of a Mathematical Model and Sensitive Analysis." *Journal of Food Engineering* 96 (1): 141–50. https://doi.org/10.1016/j.jfoodeng.2009.07.005.

Kowalski, Stefan J., and Dominik Mierzwa. 2011. "Hybrid Drying of Red Bell Pepper: Energy and Quality Issues." *Drying Technology* 29 (10): 1195–203. https://doi.org/10.1080/07373937.2011.578231.

Kumar, Chandan, M. A. Karim, and Mohammad U. H. Joardder. 2014. "Intermittent Drying of Food Products: A Critical Review." *Journal of Food Engineering* 121 (January): 48–57. https://doi.org/10.1016/j.jfoodeng.2013.08.014.

Kwaśnica, Andrzej, Natalia Pachura, Klaudia Masztalerz, Adam Figiel, Aleksandra Zimmer, Robert Kupczyński, Katarzyna Wujcikowska, Angel A. Carbonell-Barrachina, Antoni Szumny, and Henryk Różański. 2020. "Volatile Composition and Sensory Properties as Quality Attributes of Fresh and Dried Hemp Flowers (Cannabis sativa L.)." *Foods* 9 (8): 1118. https://doi.org/10.3390/foods9081118.

Laribi, Bochra, Karima Kouki, Mahmoud M'Hamdi, and Taoufik Bettaieb. 2015. "Coriander (Coriandrum sativum L.) and Its Bioactive Constituents." *Fitoterapia* 103 (June): 9–26. https://doi.org/10.1016/j.fitote.2015.03.012.

Łechtańska, J. M., J. Szadzińska, and S. J. Kowalski. 2015. "Microwave- and Infrared-Assisted Convective Drying of Green Pepper: Quality and Energy Considerations." *Chemical Engineering and Processing: Process Intensification* 98 (December): 155–64. https://doi.org/10.1016/j.cep.2015.10.001.

Liu, Yunhong, Yue Sun, Anguo Xie, Huichun Yu, Yong Yin, Xin Li, and Xu Duan. 2017. "Potential of Hyperspectral Imaging for Rapid Prediction of Anthocyanin Content of Purple-Fleshed Sweet Potato Slices During Drying Process." *Food Analytical Methods* 10 (12): 3836–46. https://doi.org/10.1007/s12161-017-0950-y.

Liu, Lijun, Yuxin Wang, Dandan Zhao, Kejing An, Shenghua Ding, and Zhengfu Wang. 2014. "Effect of Carbonic Maceration Pre-Treatment on Drying Kinetics of Chilli (Capsicum Annuum L.) Flesh and Quality of Dried Product." *Food and Bioprocess Technology* 7 (9): 2516–27. https://doi.org/10.1007/s11947-014-1253-6.

Łyczko, Jacek, Klaudiusz Jałoszyński, Mariusz Surma, José Miguel García-Garví, Ángel Antonio Carbonell-Barrachina, and Antoni Szumny. 2019a. "Determination of Various Drying Methods' Impact on Odour Quality of True Lavender (Lavandula angustifolia Mill.) Flowers." *Molecules* 24 (16): 2900. https://doi.org/10.3390/molecules24162900.

Łyczko, Jacek, Klaudiusz Jałoszyński, Mariusz Surma, Klaudia Masztalerz, and Antoni Szumny. 2019b. "HS-SPME Analysis of True Lavender (Lavandula angustifolia Mill.) Leaves Treated by Various Drying Methods." *Molecules* 24 (4). https://doi.org/10.3390/molecules24040764.

Łyczko, Jacek, Klaudia Masztalerz, Leontina Lipan, Hubert Iwiński, Krzysztof Lech, Ángel A. Carbonell-Barrachina, and Antoni Szumny. 2021. "Coriandrum sativum L.—Effect of Multiple Drying Techniques on Volatile and Sensory Profile." *Foods* 10(2): 403.

Łyczko, Jacek, Klaudia Masztalerz, Leontina Lipan, Krzysztof Lech, Ángel A. Carbonell-Barrachina, and Antoni Szumny. 2020. "Chemical Determinants of Dried Thai Basil (O. Basilicum Var. Thyrsiflora) Aroma Quality." *Industrial Crops and Products* 155 (November): 112769. https://doi.org/10.1016/j.indcrop.2020.112769.

Madhava Naidu, M., M. Vedashree, Pankaj Satapathy, Hafeeza Khanum, Ravi Ramsamy, and H. Umesh Hebbar. 2016. "Effect of Drying Methods on the Quality Characteristics of Dill (Anethum graveolens) Greens." *Food Chemistry* 192 (February): 849–56. https://doi.org/10.1016/j.foodchem.2015.07.076.

Mahendran, Ganesan, and Laiq-Ur Rahman. 2020. "Ethnomedicinal, Phytochemical and Pharmacological Updates on Peppermint (Mentha×Piperita L.)—A Review." *Phytotherapy Research* 34 (9): 2088–139. https://doi.org/10.1002/ptr.6664.

Majumder, Prasanta, Abhijit Sinha, Rajat Gupta, and Shyam S. Sablani. 2021. "Drying of Selected Major Spices: Characteristics and Influencing Parameters, Drying Technologies, Quality Retention and Energy Saving, and Mathematical Models." *Food and Bioprocess Technology* 14 (6): 1028–54. https://doi.org/10.1007/s11947-021-02646-7.

Michelino, Filippo, Alessandro Zambon, Matteo Tobia Vizzotto, Stefano Cozzi, and Sara Spilimbergo. 2018. "High Power Ultrasound Combined with Supercritical Carbon Dioxide for the Drying and Microbial Inactivation of Coriander." *Journal of CO$_2$ Utilization* 24 (March): 516–21. https://doi.org/10.1016/j.jcou.2018.02.010.

Miraei Ashtiani, Seyed-Hassan, Alireza Salarikia, and Mahmood Reza Golzarian. 2017. "Analyzing Drying Characteristics and Modeling of Thin Layers of Peppermint Leaves under Hot-Air and Infrared Treatments." *Information Processing in Agriculture* 4 (2): 128–39. https://doi.org/10.1016/j.inpa.2017.03.001.

Mujumdar, Arun S. 2007. "An Overview of Innovation in Industrial Drying: Current Status and R&D Needs." *Transport in Porous Media* 66 (1): 3–18. https://doi.org/10.1007/s11242-006-9018-y.

Nguyen, Huong Thi, and Zámboriné Éva Németh. 2016. "Sources of Variability of Wormwood (Artemisia Absinthium L.) Essential Oil." *Journal of Applied Research on Medicinal and Aromatic Plants* 4 (3): 143–50. https://doi.org/10.1016/j.jarmap.2016.07.005.

Nindo, C. I., and J. Tang. 2007. "Refractance Window Dehydration Technology: A Novel Contact Drying Method." *Drying Technology* 25 (1): 37–48. https://doi.org/10.1080/07373930601152673.

Ning, Xiaofeng, Yulong Feng, Yuanjuan Gong, Yongliang Chen, Junwei Qin, and Danyang Wang. 2019. "Drying Features of Microwave and Far-Infrared Combination Drying on White Ginseng Slices." *Food Science and Biotechnology* 28 (4): 1065–72. https://doi.org/10.1007/s10068-018-00541-0.

Nowacka, Malgorzata, and Malgorzata Wedzik. 2016. "Effect of Ultrasound Treatment on Microstructure, Colour and Carotenoid Content in Fresh and Dried Carrot Tissue." *Applied Acoustics* 103 (February): 163–71. https://doi.org/10.1016/j.apacoust.2015.06.011.

Orphanides, Antia, Vlasios Goulas, and Vassilis Gekas. 2016. "Drying Technologies: Vehicle to High-Quality Herbs." *Food Engineering Reviews* 8 (2): 164–80. https://doi.org/10.1007/s12393-015-9128-9.

Osae, Richard, Cunshan Zhou, Baoguo Xu, William Tchabo, Haroon Elrasheid Tahir, Abdullateef Taiye Mustapha, and Haile Ma. 2019. "Effects of Ultrasound, Osmotic Dehydration, and Osmosonication Pretreatments on Bioactive Compounds, Chemical Characterization,

Enzyme Inactivation, Color, and Antioxidant Activity of Dried Ginger Slices." *Journal of Food Biochemistry* 43 (5): e12832. https://doi.org/10.1111/jfbc.12832.

Öztekin, S., and Milan Martinov. 2014. *Medicinal and Aromatic Crops: Harvesting, Drying, and Processing.* https://open.uns.ac.rs/handle/123456789/12985.

Parlak, Nezaket. 2015. "Fluidized Bed Drying Characteristics and Modeling of Ginger (Zingiber Officinale) Slices." *Heat and Mass Transfer* 51 (8): 1085–95. https://doi.org/10.1007/s00231-014-1480-4.

Patel, Sanjay Kumar, and Mukund Haribhau Bade. 2021. "Superheated Steam Drying and Its Applicability for Various Types of the Dryer: The State of Art." *Drying Technology* 39 (3): 284–305. https://doi.org/10.1080/07373937.2020.1847139.

Patočka, J., and Jiří Jákl. 2010. "Biomedically Relevant Chemical Constituents of Valeriana officinalis." *Journal of Applied Biomedicine* 8: 11–8. https://doi.org/10.2478/V10136-009-0002-Z.

Phoungchandang, S., S. Nongsang, and P. Sanchai. 2009. "The Development of Ginger Drying Using Tray Drying, Heat Pump–Dehumidified Drying, and Mixed-Mode Solar Drying." *Drying Technology* 27 (10): 1123–31. https://doi.org/10.1080/07373930903221424.

Ps, Anandu, K. Sangeetha, Sanjana Potluri, R. Santhosh, and R. Mahendran. 2018. "Physical Properties of Infrared (IR) Assisted Hot Air Dried Nutmeg (Myristica Fragrans) Seeds." *Journal of Food Processing and Preservation* 42 (1): e13359. https://doi.org/10.1111/jfpp.13359.

Putsakum, Gontorn, Nurain Rahman, Hanisah Kamilah, Kaiser Mahmood, and Fazilah Ariffin. 2020. "The Effects of Blanching Pretreatment and Immersion of Sodium Metabisulfite/Citric Acid Solution on the Myristicin Content and the Quality Parameter of Nutmeg (Myristica fragrans) Pericarp." *Journal of Food Measurement and Characterization* 14 (6): 3455–61. https://doi.org/10.1007/s11694-020-00584-0.

Qiu, Liqing, Min Zhang, Arun S. Mujumdar, and Yaping Liu. 2020. "Recent Developments in Key Processing Techniques for Oriental Spices/Herbs and Condiments: A Review." *Food Reviews International* (November): 1–21. https://doi.org/10.1080/87559129.2020.1839492.

Raghavan, Susheela. 2006. *Handbook of Spices, Seasonings, and Flavorings.* 2nd ed. CRC Press. https://www.routledge.com/Handbook-of-Spices-Seasonings-and-Flavorings/Raghavan/p/book/9780367390099.

Rahman, Nurain, Tan Bee Xin, Hanisah Kamilah, and Fazilah Ariffin. 2018. "Effects of Osmotic Dehydration Treatment on Volatile Compound (Myristicin) Content and Antioxidants Property of Nutmeg (Myristica fragrans) Pericarp." *Journal of Food Science and Technology* 55 (1): 183–9. https://doi.org/10.1007/s13197-017-2883-2.

Reich, Peter B., Mark G. Tjoelker, Kurt S. Pregitzer, Ian J. Wright, Jacek Oleksyn, and Jose-Luis Machado. 2008. "Scaling of Respiration to Nitrogen in Leaves, Stems and Roots of Higher Land Plants." *Ecology Letters* 11 (8): 793–801. https://doi.org/10.1111/j.1461-0248.2008.01185.x.

Rocha, R. P. 2011. "Influence of Drying Process on the Quality of Medicinal Plants: A Review." *Journal of Medicinal Plants Research* 5 (33). https://doi.org/10.5897/JMPRX11.001.

Rodríguez, J., A. Mulet, and J. Bon. 2014. "Influence of High-Intensity Ultrasound on Drying Kinetics in Fixed Beds of High Porosity." *Journal of Food Engineering* 127 (April): 93–102. https://doi.org/10.1016/j.jfoodeng.2013.12.002.

Romdhana, Hedi, Catherine Bonazzi, and Martine Esteban-Decloux. 2015. "Superheated Steam Drying: An Overview of Pilot and Industrial Dryers with a Focus on Energy Efficiency." *Drying Technology* 33 (10): 1255–74. https://doi.org/10.1080/07373937.2015.1025139.

Saengrayap, Rattapon, Ampawan Tansakul, and Gauri S. Mittal. 2015. "Effect of Far-Infrared Radiation Assisted Microwave-Vacuum Drying on Drying Characteristics and Quality of Red Chilli." *Journal of Food Science and Technology* 52 (5): 2610–21. https://doi.org/10.1007/s13197-014-1352-4.

Sagoo, S. K., C. L. Little, M. Greenwood, V. Mithani, K. A. Grant, J. McLauchlin, E. de Pinna, and E. J. Threlfall. 2009. "Assessment of the Microbiological Safety of Dried Spices and Herbs from Production and Retail Premises in the United Kingdom." *Food Microbiology* 26 (1): 39–43. https://doi.org/10.1016/j.fm.2008.07.005.

Schössler, Katharina, Henry Jäger, and Dietrich Knorr. 2012. "Effect of Continuous and Intermittent Ultrasound on Drying Time and Effective Diffusivity During Convective Drying of Apple and Red Bell Pepper." *Journal of Food Engineering* 108 (1): 103–10. https://doi.org/10.1016/j.jfoodeng.2011.07.018.

Sehrawat, Rachna, and Prabhat K. Nema. 2018. "Low Pressure Superheated Steam Drying of Onion Slices: Kinetics and Quality Comparison with Vacuum and Hot Air Drying in an Advanced Drying Unit." *Journal of Food Science and Technology* 55 (10): 4311–20. https://doi.org/10.1007/s13197-018-3379-4.

Sehrawat, Rachna, Prabhat K. Nema, and Barjinder Pal Kaur. 2016. "Effect of Superheated Steam Drying on Properties of Foodstuffs and Kinetic Modeling." *Innovative Food Science & Emerging Technologies* 34 (April): 285–301. https://doi.org/10.1016/j. ifset.2016.02.003.

Singh, Neetu, Amrender Singh Rao, Abhishek Nandal, Sanjiv Kumar, Surender Singh Yadav, Showkat Ahmad Ganaie, and Balasubramanian Narasimhan. 2021. "Phytochemical and Pharmacological Review of Cinnamomum Verum J. Presl-a Versatile Spice Used in Food and Nutrition." *Food Chemistry* 338 (February): 127773. https://doi.org/10.1016/j. foodchem.2020.127773.

Smigielski, K., R. Prusinowska, A. Stobiecka, A. Kunicka-Styczyñska, and R. Gruska. 2018. "Biological Properties and Chemical Composition of Essential Oils from Flowers and Aerial Parts of Lavender (Lavandula angustifolia)." *Journal of Essential Oil Bearing Plants* 21 (5): 1303–14. https://doi.org/10.1080/0972060X.2018.1503068.

Sospedra, Isabel, Jose M. Soriano, and Jordi Mañes. 2010. "Assessment of the Microbiological Safety of Dried Spices and Herbs Commercialized in Spain." *Plant Foods for Human Nutrition* 65 (4): 364–68. https://doi.org/10.1007/s11130-010-0186-0.

Stępień, Agnieszka Ewa, Józef Gorzelany, Natalia Matłok, Krzysztof Lech, and Adam Figiel. 2019. "The Effect of Drying Methods on the Energy Consumption, Bioactive Potential and Colour of Dried Leaves of Pink Rock Rose (Cistus creticus)." *Journal of Food Science and Technology* 56 (5): 2386–94. https://doi.org/10.1007/s13197-019-03656-2.

Sukumaran, R., B. L. Chua, and N. Ismail. 2021. "Hybrid Drying of Ultrasound Assisted Osmotic Dehydration (UOAD) and Hot Air Drying of Eucalyptus Deglupta." *Journal of Physics: Conference Series* 2120 (1): 012003. https://doi.org/10.1088/1742-6596/2120/1/012003.

Szopa, Agnieszka, Joanna Pajor, Paweł Klin, Agnieszka Rzepiela, Hosam O. Elansary, Fahed A. Al-Mana, Mohamed A. Mattar, and Halina Ekiert. 2020. "Artemisia Absinthium L.-Importance in the History of Medicine, the Latest Advances in Phytochemistry and Therapeutical, Cosmetological and Culinary Uses." *Plants (Basel, Switzerland)* 9 (9): E1063. https://doi.org/10.3390/plants9091063.

Tan, Yuen Ping, and Eric Wei Chiang Chan. 2014. "Antioxidant, Antityrosinase and Antibacterial Properties of Fresh and Processed Leaves of Anacardium Occidentale and Piper Betle." *Food Bioscience* 6 (June): 17–23. https://doi.org/10.1016/j.fbio.2014.03.001.

Thamkaew, Grant, Ingegerd Sjöholm, and Federico Gómez Galindo. 2021. "A Review of Drying Methods for Improving the Quality of Dried Herbs." *Critical Reviews in Food Science and Nutrition* 61 (11): 1763–86. https://doi.org/10.1080/10408398.2020.1765309.

Vadivambal, R., and D. S. Jayas. 2010. "Non-Uniform Temperature Distribution during Microwave Heating of Food Materials—A Review." *Food and Bioprocess Technology* 3 (2): 161–71. https://doi.org/10.1007/s11947-008-0136-0.

Vega-Gálvez, Antonio, Elsa Uribe, Mario Perez, Gipsy Tabilo-Munizaga, Judith Vergara, Purificación Garcia-Segovia, Elena Lara, and Karina Di Scala. 2011. "Effect of High Hydrostatic Pressure Pretreatment on Drying Kinetics, Antioxidant Activity, Firmness and Microstructure of Aloe Vera (Aloe barbadensis Miller) Gel." *LWT - Food Science and Technology* 44 (2): 384–91. https://doi.org/10.1016/j.lwt.2010.08.004.

Wanas, Amira S., Mohamed M. Radwan, Suman Chandra, Hemant Lata, Zlatko Mehmedic, Abbas Ali, KHC Baser, Betul Demirci, and Mahmoud A. ElSohly. 2020. "Chemical Composition of Volatile Oils of Fresh and Air-Dried Buds of Cannabis Chemovars, Their Insecticidal and Repellent Activities." *Natural Product Communications* 15 (5): 1934578X20926729. https://doi.org/10.1177/1934578X20926729.

Weil, M., A. Shum Cheong Sing, J. M. Méot, R. Boulanger, and P. Bohuon. 2017. "Impact of Blanching, Sweating and Drying Operations on Pungency, Aroma and Color of Piper Borbonense." *Food Chemistry* 219 (March): 274–81. https://doi.org/10.1016/j.foodchem.2016.09.144.

Won, Yu-Chul, Sea C. Min, and Dong-Un Lee. 2015. "Accelerated Drying and Improved Color Properties of Red Pepper by Pretreatment of Pulsed Electric Fields." *Drying Technology* 33 (8): 926–32. https://doi.org/10.1080/07373937.2014.999371.

Yang, Fanli, Min Zhang, Arun S. Mujumdar, Qifeng Zhong, and Zhushang Wang. 2018. "Enhancing Drying Efficiency and Product Quality Using Advanced Pretreatments and Analytical Tools—An Overview." *Drying Technology* 36 (15): 1824–38. https://doi.org/10.1080/07373937.2018.1431658.

Younis, Mahmoud, Diaeldin Abdelkarim, and Assem Zein El-Abdein. 2018. "Kinetics and Mathematical Modeling of Infrared Thin-Layer Drying of Garlic Slices." *Saudi Journal of Biological Sciences* 25 (2): 332–8. https://doi.org/10.1016/j.sjbs.2017.06.011.

Zambon, Alessandro, Filippo Michelino, Siméon Bourdoux, Frank Devlieghere, Stefania Sut, Stefano Dall'Acqua, Andreja Rajkovic, and Sara Spilimbergo. 2018. "Microbial Inactivation Efficiency of Supercritical CO_2 Drying Process." *Drying Technology* 36 (16): 2016–21. https://doi.org/10.1080/07373937.2018.1433683.

Zhang, Bin, Zhichang Qiu, Ruixuan Zhao, Zhenjia Zheng, Xiaoming Lu, and Xuguang Qiao. 2021a. "Effect of Blanching and Freezing on the Physical Properties, Bioactive Compounds, and Microstructure of Garlic (Allium sativum L.)." *Journal of Food Science* 86 (1): 31–39. https://doi.org/10.1111/1750-3841.15525.

Zhang, Mengmeng, Rong Zhao, Dan Wang, Li Wang, Qing Zhang, Shujun Wei, Feng Lu, Wei Peng, and Chunjie Wu. 2021b. "Ginger (Zingiber officinale Rosc.) and Its Bioactive Components Are Potential Resources for Health Beneficial Agents." *Phytotherapy Research: PTR* 35 (2): 711–42. https://doi.org/10.1002/ptr.6858.

9 Techno-Economic Evaluation for Cost-Effective Drying of Herbs, Spices and Medicinal Plants

Viplav Hari Pise and Bhaskar N. Thorat
Institute of Chemical Technology

CONTENTS

9.1 INTRODUCTION

Drying is a unique process for preserving agricultural produces, medicinal herbs, and aromatic plants. Drying lowers the moisture content reducing microbial and enzymatic activity. Along with improved shelf-life, it also reduces the density and decreases the transport cost. The dehydration process needs to be suitably selected from the existing technologies so that water activity is adjusted to a level where microbial activity is the least from a preservation point of view. The dehydration process should be able to retain the characteristics of the product from the application point of view (Thamkaew et al., 2021). Low-moisture products typically have a moisture content of <25% and water activity between 0.0 and 0.60, and intermittent-moisture products have a moisture content between 15% and 50% and water activity between 0.60 and 0.85 (Taoukis & Richardson, 2007). The most desired purpose of dehydration is to reduce the moisture

DOI: 10.1201/9781003269250-9

present in herbs, spices or other parts of medicinal plants without affecting their key attributes for further use (Bhaskara Rao & Murugan, 2021). Dehydration takes place by the application of heat and mass transfer at the cellular level. Quantification and ways of removing free water and inter- and intra-cellular water need to be understood through experimental and mathematical analysis (Khan et al., 2017). The detailed analysis regarding dehydration at the cellular level is not only valuable for determining the parameters affecting drying, including the activation energy, driving force/concentration gradient, internal and external mass transfer rate and effective moisture diffusion (Majumder et al., 2021) but also it will help in the understanding of conditions required for the retention of desired phytochemicals and volatiles in herbs, spices and aromatic plants (Pise et al., 2022; Prothon et al., 2003). It is also highly desired to obtain the critical parameters for energy conservation vis-à-vis the desirable thermal conditions (Rahman et al., 2018). These parameters can be controlled by mode of energy application, temperature, flow/draft and relative humidity of the drying medium and the size of the material being dried. Advances in drying technology through different dehydrators and solar dryers for agro-commodities, food applications (Calín-Sánchez et al., 2020; Radoj et al., 2021; Uthpala et al., 2020), herbs and spices (Bhaskara Rao & Murugan, 2021; Jin et al., 2018; Majumder et al., 2021; Orphanides et al., 2016; Qiu et al., 2020; Thamkaew et al., 2021) have been well summarised. Solar applications for dehydration have been specifically looked into in these reviews, considering the cost benefits based on the commercial fuel savings (Bhaskara Rao & Murugan, 2021; Chavan et al., 2020; Chavan & Thorat, 2020, 2021; Kamarulzaman et al., 2021). An efficient dehydration technique specific to a given application can be worked out by necessary modifications in the design. Such alterations as per the region-specific conditions, local produces and harvest seasons can significantly improve the ecosystem, besides an impactful socio-economic changeover (Orsat et al., 2008). This is true in the case of upcoming and developing economies of the world, which are struggling to create a livelihood for a sizeable chunk of the population. This chapter highlights selection criteria in adopting the most suitable technology, under varying operating conditions. The techno-economic evaluation of these dryers will certainly benefit the stakeholders that include the peasants and self-help groups. The tribal people would be another beneficiary who would get involved in the local and primary industrial activity based on 'non-timber forest products'. As one can easily see, several South and Southeast Asian countries are endowed with the forest area and the dependent tribal population.

9.2   HERBS, SPICES AND MEDICINAL PLANTS – IMPORTANCE AND NEED OF DEHYDRATION

The usage of herbs and spices has been known for over 2,000 years. Herbs and spices had long been considered of economic importance for their medicinal applications, preservative properties and flavour, aroma and colour contributing to culinary applications. They are also used in preservatives considering their powerful anti-oxidant properties. A herb, in botanical terms, is 'any plant with a soft succulent tissue'. On a broader aspect, herbs can be classified as leafy products, whereas, spices come as any part of the plant like bud, flower, bark, root, fruit/berry or seed. They are also classified as 'temperate zone origin' and 'tropical aromatic', respectively (Pearson &

Gillett, 1996). Though this classification is not clearly defined, key characteristics of these medicinal or aromatic herbs and spices are due to the presence of volatile oils and oleoresins in their parts. Phytochemicals like polyphenols (phenolic acids, flavones and flavonols), terpenoids and alkaloids in synergy with minor and trace compounds of the matrix are the essential compounds contributing to the properties and desired effects. Potent anti-oxidant and anti-microbial (anti-bacterial, antiviral and anti-fungal) properties promote immunity and therapeutic applications. Sedative, stress-relieving and calming properties act as nerve tonics, thus promoting overall health. Anti-inflammatory, anti-septic, anti-spasmodic, carminative and stomachic properties of these herbs and spices lead to applications towards digestive, respiratory, blood circulatory and muscle systems (Embuscado, 2015; Lawless, 1992; Opara & Chohan, 2014; Vázquez-Fresno et al., 2019). Hence, the preservation of these compounds in the form of volatile oils is important. These are best preserved in their natural matrix, which makes dehydration of these a critical operation.

TABLE 9.1
Herbs and Spices – Water Contents and Volatiles

Sr. No.	Herbs and Spices	Part of Plant	Initial Moisture Content % w.b.	Desired Moisture Content % w.b.	Essential Oils	Oleo-Resins	References
1	Anise	Seed	15–24.8	~10	2.9	15.4	Kürkçüoğlu et al. (2007), Leal et al. (2011) and Yamini et al. (2008)
2	Basil	Leaves	75–80	~15	0.21–0.5	NA	Hossain et al. (2010) and Raina et al. (2013)
3	Black pepper	Fruit/ corn	81.3	10–11	1.1	5.6	Majumder et al. (2021) and Pearson and Gillett (1996)
4	Caraway	Seeds	22–24	~10	2.8	6	Pearson and Gillett (1996)
5	Cardamon	Fruit/ seeds	24	~10	1.5	7.6	Marongiu et al. (2004) and Pearson and Gillett (1996)
6	Celery	Leaves	86–91	10–12	0.6	0.9	Sellami et al. (2012)
7	Celery seeds	Seeds	5.2–20.2	4–5	2	5.5	Pearson and Gillett (1996)
8	Chilli pepper	Pods	75–83	10–11			Majumder et al. (2021)

(Continued)

TABLE 9.1 (*Continued*)
Herbs and Spices – Water Contents and Volatiles

Sr. No.	Herbs and Spices	Part of Plant	Initial Moisture Content % w.b.	Desired Moisture Content % w.b.	Essential Oils	Oleo-Resins	References
9	Cinnamon	Bark	60	14	1.3	6.83	Kamaliroosta (2012)
10	Clove	Flower bud	28–30	13	10.1	34.3	Guan et al. (2007)
11	Coriander	Leaves	85–87	13–14	0.6	NA	Sourmaghi et al. (2015)
12	Coriander seeds	Seeds	18	8–12	0.7	3	Pearson and Gillett (1996)
13	Cumin	Seeds	18–22	8–12	3.16	15.6	Benmoussa et al. (2018)
14	Dill seeds	Seeds	22–24	8–12	3.2	4.5	Pearson and Gillett (1996)
15	Fennel	Seeds	62–65	10–12	0.98	2.2	Hammouda et al. (2013)
16	Fenugreek	Seeds	22–24	~10			Balasubramanian et al. (2016)
17	Ginger	Rhizome	85–90	8–10	0.3	3.5	Majumder et al. (2021) and Pearson and Gillett (1996)
18	Mace	Aril			5	7	Pearson and Gillett (1996)
19	Marjoram	Leaves	75–80	~15	0.4	2.2	Hossain et al. (2010) and Pearson and Gillett (1996)
20	Nutmeg	Seed			5.5	7.5	Hossain et al. (2010) and Pearson and Gillett (1996)
21	Oregano	Leaves	75–80	~15	2.5–5	NA	
22	Parsley	Leaves	78–82	10–11	3.5–4	NA	
23	Parsley seeds	Seeds		8	2.9	4.5	Pearson and Gillett (1996)
24	Peppermint	Leaves	82–85	11	1.1	1.5	Babu et al. (2018) and Pearson and Gillett (1996)

(Continued)

TABLE 9.1 (*Continued*)
Herbs and Spices – Water Contents and Volatiles

Sr. No.	Herbs and Spices	Part of Plant	Initial Moisture Content % w.b.	Desired Moisture Content % w.b.	Essential Oils	Oleo-Resins	References
25	Rosemary	Leaves	75–80	~15	0.35	NA	Bousbia et al. (2009) and Hossain et al. (2010)
26	Saffron	Stigma	75–77	8–10	0.41	1.31	Shao et al. (2014)
27	Sage	Leaves	75–80	~15	2	4	Hossain et al. (2010) and Pearson and Gillett (1996)
28	Spearmint	Leaves	82–85	11	0.75	NA	Babu et al. (2018) and Shahi et al. (1999)
29	Tarragon	Leaves					
30	Thyme	Leaves	75–80	~15	2.39	NA	Golmakani and Moayyedi (2015) and Hossain et al. (2010)
31	Turmeric	Rhizome	83–87	11–12	0.45–1.2	5.49	Majumder et al. (2021) and Manzan et al. (2003)
32	Garlic	Bulb	65–70	6–7			Majumder et al. (2021)
33	Onions	Bulb	86–90	4–6			Majumder et al. (2021) and Savitha et al. (2022)

The initial moisture content in some of the herbs, desired moisture content for better shelf life, and retention of the volatile oil content can be seen in Table 9.1. Almost all herbs are seasonal; hence, drying becomes crucial for the ease of mobility, consumption and availability throughout the year as per the consumer's demand. Even at small-scale consumption, a high value associated with dried herbs generally leads to high expectations of consumers on quality (Thamkaew et al., 2021). Quality parameter of herbs and the herbal finished product is closely linked with unit operations of pre-harvesting, harvesting and processing (drying, storage and extraction)

(Tanko et al., 2005). As drying is the first unit operation in the process with controllable parameters, it plays a crucial role in retaining quality. Drying typically leads to decolourisation, alteration in textures and modification in the composition and quality of raw material if not carried out correctly. Hence, drying is extensively explored to meet market expectations.

Food, pharmaceutical, cosmetic, flavour and fragrance industries are increasingly interested in natural herbs, spices and medicinal plants (Hossain et al., 2010). Almost 184 and 222 countries are exporting and importing, respectively, spices worth about USD 2.879 billion, accounting for 0.016% of the world's trade. China ($737M), India ($388M), Netherlands ($195M), Germany ($124M) and Turkey ($112M) are the major exporters and the United States ($307M), Netherlands ($169M), Germany ($159M), Saudi Arabia ($129M), and the United Kingdom ($120M) are the major importers (Simoes & Hidalgo, 2021). Hence, exploration of drying principles, advancement of technology and implementation through drying units to maintain the goodness of these herbs, spices and medicinal plants is needed. Considering the market volume, production cycles, industrial requirements, equipment sizing, process conditions and associated costs needs to be studied and optimised to make the commercial process feasible.

9.3   DRYING – PARAMETERS AND OPTIMISATION

Drying at the molecular level involves activating water molecules, mobilising water molecules within the matrix and transferring molecules outside the matrix. These three activities are carried out sequentially and at times simultaneously in dryers by controlling parameters such as the temperature of drying for the activation energy, mode of heating for internal diffusion and draft (natural or forced) and surface area exposure for external mass transfer (Khan et al., 2017; Mujumdar, 2014). The technical performances of the dryer, including drying rate (DR), capacity, energy consumption and efficiency, can be regulated, as required, by controlling each of these parameters. Different drying techniques and the combination thereof have evolved, considering the products to be dried, and desired responses were colour, shrinkage, bulk density, porosity, phytochemical and volatile retention, anti-oxidants, sugars, proteins and sensory attributes. An ideal dried product should reconstitute all properties of the original product on rehydration (Khaing Hnin et al., 2019).

In natural drying, mass transfer through the desiccating effect of air and activation energy for the evaporation of water molecules by exposure to the sun supports the process. Exposure to the sun can be controlled and enhanced by using different configurations of solar dryers. In dehydration operation, the temperatures are maintained in the drying chamber from freezing to hot conditions, i.e. −50°C to 100°C for herbs and spices, using refrigerated systems and heaters (fuelled by electric, natural gas, coal/briquettes, firewood and so on), respectively. Considering the size, shape and structure of the objects, this temperature or the activation energy can be supplied through direct or indirect heat, such as conduction, convection or radiation mode of heat transfer. There are other means of drying such as microwave (MW), far-infrared, infrared, ultra-violet and radiofrequency energy. The external mass transfer conditions are controlled by circulation draft (natural or forced), vacuum (regulating partial pressures) and relative humidity for providing moisture gradient.

Regulating these parameters, the energy of activation for vaporisation and the inherently slow process of drying makes this unit operation one of the most energy-intensive (Mujumdar, 2014).

Having said this, the product's quality is closely related to these dehydration parameters as seen at the micro-structural level (Savitha et al., 2022). The rate at which the moisture is removed must be well-balanced to control the quality of the dried product. High-temperature drying and the most significant variation between the internal diffusion and evaporation rate can be catastrophic, causing permanent physical and chemical changes (Tanko et al., 2005). Herbs are leafy biomasses, having a tender structure, high moisture content, thin layer and high sensitivity towards temperature due to the presence of flavour, nutrients, colour and texture of interest (Babu et al., 2018; Hossain et al., 2010). During the drying operation, the rate of dehydration can be governed by either internal mass diffusion or external mass transfer. A high temperature leads to faster water removal leading to unbalanced internal moisture diffusion and external mass transfer. However, due to the quicker mass transfer rate, the internal stresses at the cellular level are seen to increase, resulting in the loss of volatiles (Pise et al., 2022; Savitha et al., 2022). The novel approach defined, based on the drying, by Pai et al. identifies the control over dehydration by knowing the relative rates of internal diffusion and external mass transfer rate (Chavan & Thorat, 2020; Pai et al., 2021). For ideal drying conditions, the controlled activation energy for balanced internal mass diffusion and external mass transfer, low temperature and high surface areas are required, which may lead to the higher cost of the drying system. The most suitable dryer offers optimised conditions for higher capacity operation, better quality products, good efficiency, lower cost and minimal environmental impact (Khaing Hnin et al., 2019). From a research point of view, the drying process mainly depends on the product size (microns to tens of centimetres), product porosity (0%–99%), drying time (0.25 seconds to 5 months), production capacity (0.1 kg/h to 100 tons/h), product speed (stationary to 2,000 m/min) or the residence time, drying temperature (below triple point to above the critical point of water), operating pressure (fraction of millibar to 25 atm) and mode of heat transfer (Mujumdar, 2014). Thus, it can be seen that the drying process needs a great deal of attention. The right choice of the most suitable dryer must provide a good return on investment (RoI) and is techno-economically the most feasible.

9.4 NEED FOR TECHNO-ECONOMIC EVALUATION OF DRYING PROCESS

The selection of dryers considering criticalities as mentioned in the above section plays an important role in preserving the produce, retaining quality, sustainability of the process, economic viability and contribution to the overall growth and development through the manufacturing activity. A detailed techno-economic feasibility study (TEFS) ensures proper dryer selection, safeguards the investment and prevents losses, both tangible as well as intangible (Subramanian & Taghizadeh-Hesary, 2021).

The risk of failure in implementing the dryer depends on intrinsic and extrinsic factors. *Intrinsic factors* like the optimum performance of the dryer, operating parameters, application energy and its mode, loading capacity and DR determine the quality of the products, and extrinsic factors like the ease of operation, utility and workforce requirement, duration of the batch, capital cost and annual operational days ensure the profitability of the unit. TEFS for any dryer determines the risks associated with units and highlights the areas of concern to be worked upon to meet the economic viability and the financial feasibility of the project. The profitability aspect of the proposed drying project needs to be evaluated through TEFS.

TEFS helps establish a new drying system merging advancements or modifications of an existing system, contingency and mitigation plans and optimising the time frame for erection, execution, and operation for investment recovery (Subramanian & Taghizadeh-Hesary, 2021). General TEFS considers all relevant factors for implementing the project, including technical, economic, legal, scheduling and operational aspects. In this chapter, for the dehydration process, the technical elements, operational factors and financial analysis for the proposed system shall be defined to determine the feasibility of dehydration of herbs, spices and medicinal plants. A generic platform shall be provided with three case studies to guide how a TEFS is conducted. The platform can be utilised for any crop, with the valuation of the project and operational costs then, to determine the TEFS and later generate a *detailed project report* (Figure 9.1).

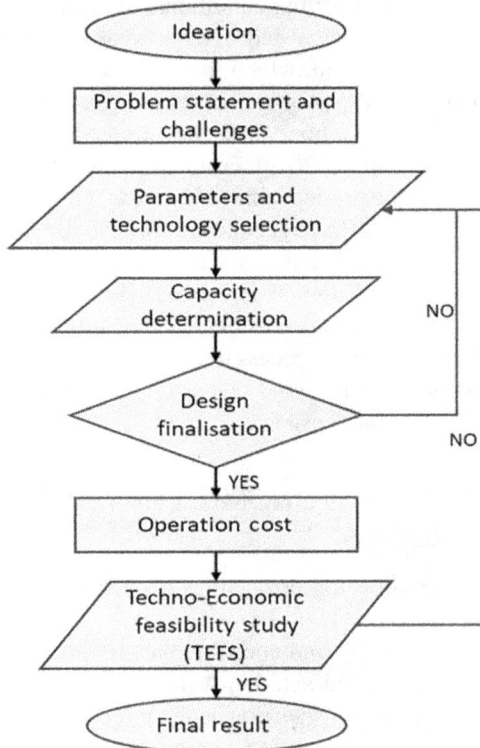

FIGURE 9.1 Dryer selection process based on the techno-economic feasibility study (TEFS).

The preparation of TEFS begins with the idea of a problem statement. The need for dehydration and the challenges associated with the process need to be clearly defined. It is then expected to overcome the processing challenges, finalise technical parameters and select the most suitable technology considering the advancements in the sector of dehydration. The availability of resources, marketable products, duration of harvest, duration of dehydration and capacity are taken into account to get the cost estimation of the unit. The operation cost along with other associated costs is taken into account. Finally, TEFS is projected based on the value addition seen through the process.

9.5 A COST-EFFECTIVE APPROACH FOR DRYING HERBS, SPICES AND MEDICINAL PLANTS

Based on the parameters mentioned in the above section for drying and its applications, these dryers can be broadly classified with a mode of heating, temperature and pressure of operation and handling of the raw material. With this classification and different product applications, more than 200 types of commercial dryers are reported in the literature. Drying of herbs itself leads to an array of dryers considering the uniqueness of each herb owing to its water content, phytochemical composition and morphological characteristics (Orphanides et al., 2016). Several dryers are available for the drying of herbs and spices classified based on the modes of heat transfer, raw material handling capacities, particle shapes, structures and sizes, time for dehydration, utility requirement, energy efficiency and capital and operating cost. Thermal exposure for the removal of moisture to yield a low moisture-containing product makes it one of the most energy-intensive unit operations (Mujumdar, 2014). Inappropriate dryers can make the process energy inefficient due to thermal energy consumption if the TEFS is not worked out properly. The feasibility of the project is typically determined and verified on multiple aspects. A few elements for conducting the studies relevant to the drying projects include technical, economical, scheduling, legal/ethical issues, resource planning, marketing, real estate and cultural and comprehensive feasibility (Subramanian & Taghizadeh-Hesary, 2021). All these need to be taken into account to ensure that the dehydration project is overall techno-economically feasible. As far as drying of herbs, spices and medicinal plants is concerned, the following three are the major focussed areas:

   i. engineering design/technical performance,
  ii. operational/functional performance and
 iii. economic analysis.

Convective heat transfer is favourable for herbs and spices, considering the high volumes and surface area for external mass transfer. Convection ensures a uniform temperature distribution and provides a draft for the initial constant rate drying phase or the surface moisture removal, especially for the herbs where the tender structure, high moisture content and thin layer of the leaves can be effectively dried. Considering the above factors, air suspension dryers, such as fluid bed type (Kafshgary et al., 2014), cyclone or rotary, air impingement (Wang et al., 2015;

Xiao & Mujumdar, 2015); packed bed or throughflow; and conveyor-truck-tunnel may be considered. The selection of a heating mode for the dehydration of these herbs and spices is crucial, considering the temperature sensitivity and the presence of volatiles and oleoresins. Multiple studies have been reported on the comparison of drying herbs using different modes of heating and retention of desired components (Calín-Sánchez et al., 2020; Jin et al., 2018; Majumder et al., 2021; Orphanides et al., 2016; Qiu et al., 2020; Thamkaew et al., 2021). Drying of these is recommended to be at a lower temperature. Vacuum conditions or negative pressure conditions can lower the temperatures or use a heat pump for the desired moisture removal from such heat-sensitive materials. In addition, parameters that affect the dryer's performance also include climatic conditions, air temperature and relative humidity (Babu et al., 2018).

As discussed above, after ideation and based on the definition of problem statements and challenges, preliminary data need to be gathered or generated through experimentation to determine the technical performance. Selection of herbs, spices or medicinal plants, followed by the listing of the initial and desired moisture content and required drying time, should be noted. Drying time is closely linked with parameters such as the loading density, draft velocity, inlet and outlet air temperature, relative humidity and airflow rate. The material thickness, density porosity and other properties also have a big influence on the drying time. These are to be monitored and mapped critically as they influence the efficiency of a scaled-up dryer. A DR curve shall be obtained for bench-scale, pilot level, demo unit and if possible plant scale drying unit for monitoring the performance (Genskow, 1994). The study of drying kinetics is an indispensable part of a drying operation or a dryer design. The conventional approach to studying the drying kinetics involves the study of a characteristic drying curve for a material encompassing a shorter constant rate period and a relatively longer falling rate period for most agricultural commodities. The empirical and semi-empirical models lay the foundation of these studies; however, they fail to provide the significance of any of the drying parameters. The regime theory derives the concept of the controlling step as the rate-determining step or the slowest step in the drying operation. Pai et al. (2021) have recently re-engineered the concept of mass transfer concept to systematically classify the drying process into different regimes as given below:

- *Regime 1* – governing regime when the drying process is controlled by the rate of internal diffusion of moisture from the core of the biomass to its surface.
- *Regime 2* – when the mass transfer from the surface of the biomass to the bulk of the environment is the controlling step.
- *Regime 3* – the drying process where both the internal diffusion of liquid and the external mass transfer process occur at a comparable rate.

Each regime is based on certain assumptions and has typical rate equations with a different set of characteristic parameters. The rate of drying in *Regime 1* shows a direct proportionality with the product of moisture present inside the sample and the corresponding holdup volume of moisture within the sample raised to a *power* (n) analogous to the order of reaction in heterogeneous chemical reaction theory. The reaction engineering approach defines the activation energy required for the drying of samples; likewise, in the mass transfer regime approach, the activation energy for *Regime 1* is given by the following equation:

$$-m_s \frac{dX}{dt} = \kappa \times \left(\rho_C \times \varepsilon_l \right)^n$$

$$\Delta E = -RT_s \ln \left(\frac{-m_s \dfrac{dX}{dt}}{\rho_c^n} \right)$$

where $-m_s \, dX/dt$ is the drying rate (kg/s), κ is the constant of Regime 1, ρ_c represents the moisture concentration at the core of the biomass, ΔE is the activation energy of moisture evaporation, R is the universal gas constant, T_s temperature of the surface and ε is the holdup volume of moisture.

It has been observed that most of the agricultural material with a dense matrix follows *Regime 1*, and alteration in different geometric shapes and drying conditions can lead to both *Regime 1* and *2* or only *Regime 2* operation. *Regime 2* involves the drying operation governed by the external mass transfer of moisture from the surface to the bulk of the dryer chamber. The rate equation and the corresponding activation energy for *Regime-2*-governed processes are given by the following equations:

$$-m_s \frac{dX}{dt} = h_m A \left(\rho_{v,s} - \rho_{v,b} \right)$$

$$\Delta E_2 = -RT_s \ln \left(\frac{-m_s \dfrac{dX}{dt} \dfrac{1}{h_m A} + \rho_{v,b}}{\rho_{v,sat}} \right)$$

Here, h_m represents the external mass transfer coefficient, A is the area over which the exchange of moisture occurs and $\rho_{v,s}$, $\rho_{v,sat}$ and $\rho_{v,b}$ are the moisture content at the surface of the biomass, at the saturated condition and the bulk of the drying chamber (kg/m^3) (Pai et al., 2021).

The knowledge of the governing regime enables one to design the drying operation; thus, it is essential to determine the regime controlling the drying process. Determination of the governing regime can be done by the following ratio (r')

$$r' = \frac{-m_s \dfrac{dX}{dt}}{h_m A \left(\rho_{v,s} - \rho_{v,b} \right)}$$

This ratio enables one to distinguish between different regimes. The value of a ratio less than 1 indicates the internal diffusion controlling process, categorised as *Regime 1*, whereas a value greater than 1 signifies *Regime 2*, external mass transfer-controlled process. With highly compact structures and densely packed ones, the occurrence of *Regime 1* could be anticipated due to the high resistance of internal diffusion to moisture. Parameters of *Regime 1* are dependent on both intrinsic and extrinsic drying conditions. Determination of parameters of *Regime 1* is done based on the regression analysis. An increase in *Regime 1* constant can be observed with an increase in the temperature and velocity while a decrease is observed with an increase in thickness and relative humidity. This suggests the *Regime 1* constant be a function of both drying conditions and the nature of biomass dried. The use of the mass transfer regime approach thus can enhance the design of the drying process.

The latent heat of vaporisation and moisture extraction rate (MER) shall be considered as mentioned in Table 9.2 to determine the heat utilised for evaporating water. Energy (utility) consumption can be determined based on the unit's thermal efficiency. This utility could be different for different dryers: for solar-based dryers – radiation received (with no additional cost), for vacuum dryer – electricity for the vacuum pump and the heating element, for heat pump dryer – electricity for dehumidification, for hot air dryer – electricity for the heating element or the units of LPG/briquettes/coal consumed and so on. Additional power requirements for the generation of the draft shall be accounted for separately. The total energy consumed shall be the sum of utility consumed and power for the draft.

Operational performance defined for the economic model shall be specific to the herbs, spices, and medicinal plants under consideration. The drying performance and conditions for leaves, barks, flowers, pods, seeds, stems, roots, rhizomes or whole plants need different considerations leading to variation in the design. Leafy plants/herbs have high moisture content immediately after harvesting. The fresh herbage of these herbs comes in huge quantities, making storage a crucial operation. If not stored correctly, spoilage of the biomass is seen on a huge scale owing to microbial growth or enzymatic or biochemical changes. The loading density, DR and operating conditions from the *technical performance* lead to the dryer chamber selection. Harvest load and feasible time duration from fresh cut to dehydration provide the required capacity of the dryer. The harvest area under cultivation gives the raw material availability or drying load on the system.

Further information shall be determined based on the details like optimum harvest per unit area, harvesting period, the season of harvesting, technical performance of the dryer and processing requirement for a season. The number of drying batches per day (residence time in the case of continuous operation) and dependent utility requirements will give the operating costs for the process. A number of operational days, the feasibility of 24-hour operation and multi-products with different harvest seasons should be opted to optimise the capacity of the dryer and distribution of the associated capital costs. Dryer selection for multiple herbs and spices with different harvest seasons can ensure a year-long operation for better cost of recovery or RoI (Table 9.3).

It also provides us with the outcome of the dried product with which the economic feasibility of the process can be determined.

TABLE 9.2

Details of Technical Performances Considered in Techno-Economic Feasibility Studies

Sr. No.	Description	Notation	Unit	Remarks
A		**Technical Performance**		
1	Initial moisture content	M_i	%	The moisture content of the fresh harvest
2	Desired moisture content	M_f	%	Safe moisture content limit for better shelf-life. Typically <20%
3	Drying time	T	h	Ideal batch time considered for achieving the desired moisture content
4	Loading capacity	B	kg	Ideal batch size considered for calculating MER and SMER
		B_{fr}	kg/h	Flow rate consideration for a continuous process, as per dehydration time
5	Drying rate	DR	kg/h	Produced dried product per unit time by the drying system
6	Water evaporated	M_w	kg	Amount of water removed by the drying systems for a batch
7	Moisture extraction rate	MER	kg/h	Water removal rate through the unit during the drying process
8	Air flow rate	M_a		Draft of air required for carrying the necessary dehydration
9	Latent heat of vaporisation	L_v	kJ/kg	
10	Inlet temperature	T_{Ai}	°C	The inlet temperature of air draft used for dehydration
11	Outlet temperature	T_{Ao}	°C	The outlet temperature of air draft used for dehydration
12	Heat utilised to evaporate water	Q_e	kJ	Based on the flow rate, specific heat and temperature of inlet and outlet air
13	Energy (utility) consumption	Q_h	kJ	Depending on the type of the dryer, mode of heating and drying conditions, energy supplied will vary
	LPG	$Q_{h(LPG)}$	kg/h	
	Biomass	$Q_{h(Biomass)}$	kg/h	
	Electricity	$Q_{h(Electricity)}$	Units	
14	Energy efficiency of dryer	$\eta_D = Q_e/Q_h$	%	
15	Heat energy requirement	Q_h	kJ	
16	Power consumption (blower)		kW	Required for providing the draft for moisture removal
17	Power factor of the motor	F		A typical value is 0.75

(*Continued*)

TABLE 9.2 (*Continued*)
Details of Technical Performances Considered in Techno-Economic Feasibility Studies

Sr. No.	Description	Notation	Unit	Remarks
18	Specific heat energy consumption	$Q_s = Q_t/M_w$	kJ/kg	
19	Specific power consumption	$Q_p = fPt/M_w$	kJ/kg	
20	Total energy requirement	$Q_t = Q_h + fPt$	kJ	

TABLE 9.3
Details of Operational Performances Considered in Techno-Economic Feasibility Studies

Sr. No.	Description	Notation	Unit	Remarks
B			**Operational Performance**	
1	Satisfactory annual harvest	Y_a	Tons/ha/year	Based on data and experience, the max/optimum/min yield per unit area can be obtained. This forms the basis for calculation
2	No. of harvest	N	Per year	Tenure of the crop and harvest season for the crop gives us the frequency of the processing load
3	Yield per harvest	Y_h	Tons/ha	For short-term crops, each harvest will give us the peak load associated with dehydration
4	Harvest seasons	t_h	Days	A number of days associated with the harvesting process at maturity. This is the minimum duration of the operation for the dryer
5	Area under consideration	A_{cap}	ha	Design/standard processing capacity for the dryer shall be peak load. Based on the same, area to be brought under cultivation to ensure smooth functioning of the dryer and uninterrupted feed will be determined
6	Processing requirement per day	C_{pro}	Tons/day	With the peak processing load determined, to minimise the losses, capacity to be processed every day needs to be finalised
7	Initial moisture content	M_i	%	The moisture content of the fresh harvest
8	Final moisture content	M_f	%	Desired moisture content post-dehydration for preservation/processing

(*Continued*)

TABLE 9.3 (*Continued*)

Details of Operational Performances Considered in Techno-Economic Feasibility Studies

Sr. No.	Description	Notation	Unit	Remarks
9	Drying time	t_d	h	Based on the drying rate curve and scale-up of the implemented drying system, the batch time along with loading/unloading time shall be finalised
10	Loading density	L_d	kg/m²	Loading on trays/conveyor/packed columns/fluidised state based on the experimental readings will give us loading density to work out chamber volume and dryer details
11	Operating condition	–	Continuous/ batch	Continuous operation – capacity processing per hour based on the required residence time shall be calculated. Batch operation – based on the drying time and the number of feasible batches, the capacity requirement of the dryer shall be calculated
12	Fresh biomass cost	C_{fresh}	USD/kg	Raw material costing is taken into account to determine the value addition through drying
13	Dried biomass cost	C_{dried}	USD/kg	Raw material cost is distributed over the reduced weight of the product and with added processing cost per unit weight

9.6 PARAMETERS FOR TECHNO-ECONOMIC EVALUATION

For an industrial process, the techno-economic evaluation of these dryers is very critical. Some dryers may be energy-efficient but have high capital costs, some are efficient in delivering the product quality but have high operating costs and some may be associated with low capital and operating costs but may require more time for dehydration leading to spoilage of the overall production. Hence, comprehensive techno-economic evaluation to meet the desired drying is critical. This brings us to the economic analysis of the dryer. The economic analysis considers the capital costs, associated financial costs (construction, erection and commissioning), operating and maintenance costs, depreciation and taxes for the system, utility and workforce costs and so on.

Techno-economic analysis, by definition, is the process of assessing the value of a given technology to guide investment and resource allocation. Techno-economic evaluation is carried out by comparing the levelised cost of processing (Kobos et al., 2020). It is the per-unit cost of drying/production over the economic lifetime for the dryer or any other technologies.

The cost of the dryer depends on the technical performance and the drying requirement, including DR or MER (also considered as throughput rate) or specific moisture removal rate (SMER, also referred to as moisture removal per unit energy consumption) and has been used in determining the dryer efficiency (η_D) (Obeng-Akrofi et al., 2021).

$$DR = \frac{(M_I - M_F)}{t}$$

where DR – drying rate (kg/h), M_I – initial moisture content (% wet basis), M_F – final moisture content (% wet basis),

$$MER = \left[W_I \left\{ 1 - \left(\frac{(1-M_i)}{(1-M_F)} \right) \right\} \right] \bigg/ t$$

where MER – moisture extraction rate (kg water/h), W_I – initial weight of the material to be dried (kg),

$$\text{SMER} = \left[W_I \times \left(\frac{(M_I - M_F)}{(100 - M_F)} \right) \right] \bigg/ Q_H$$

where SMER – specific moisture extraction rate (kg water/kWh) and Q_H – total energy consumed (kWh),

$$\eta_D = \frac{Q_E}{Q_T} \times 100 = \frac{M_a(C_iT_i - C_oT_o)t_d}{Q_H + fP_Et_d} \times 100$$

where η_D – efficiency of the dryer (%), M_a – mass flow rate of air (kg/h), C_i and C_o – specific heat capacity of inlet and outlet air (kJ/kg), T_i and T_o – inlet and outlet temperatures (°C), P_E – power consumption (kW), t_d – dehydration time (h), f – power factor of electrical motor.

With these *technical performances* obtained in terms of DR, MER, SMER and η_D, the operational cost associated with the dryer can be calculated. The efficiency of the dryer and the ratio of the energy used to evaporate the moisture from the product to the energy provided by the drying air are the determining factors for the associated operating cost. Based on the type of dryers, the utilities required can be quantified. In the case of forced convection-based solar dryers, only the power supplied for the draft is only accountable. The high energy requirement is met by increasing the collector areas, thus impacting the capital cost of equipment directly (Purohit & Kandpal, 2005). In the case of a hot air dryer, the thermal energy requirement based on the fuel (electricity/coal/LPG/briquettes) consumption along with the blower energy is to be accounted (Soysal & Öztekin, 2001). For a heat pump dryer, the energy requirement depends on the heat rejection in the condenser and the coefficient of the heat pump along with SMER and MER (Patel & Kar, 2012). If a hybrid solar dryer is considered, the assisting medium and the associated utilities are to be accounted for in determining the total energy (Fudholi et al., 2015). In addition to the utility consumption, the required workforce and associated labour costs are added. These distributed costs over the processing capacity give us the drying cost per kg of dried product.

The *economic evaluation* can be carried out based on the annualised cost method and life cycle saving. In the case of annualised cost method, the total associated drying expenses are distributed over the amount of product dried over the year. In the case of conventional energy utilisation, the drying cost keeps increasing with an increase in utility costs. However, in the case of solar dryers, the variation in the cost of drying is negligible over the operational life. Hence, to carry out the economic analysis of solar-based dehydration, the savings over the dryer's life are taken into account. In either case, the short payback period is more attractive to drive the buyers to the dehydrating process (Chavan & Thorat, 2021). The economic evaluation of the drying system takes into consideration factors like the initial cost (C_I) or capital cost of the drying unit. Details like foundation preparation, structural erection and associated auxiliary costs are also included in capital cost. Repair and maintenance cost ($C_{R\&M}$), typically 5% of the initial cost annually and depreciation cost (C_D) at a depreciation rate of about 10% to 12% annually, give the drying system's actual cost value. Scrap value (C_S), about 10% of the initial cost, is taken into account to associate the depreciation rate and the life of the equipment.

Considering C_I, C_D, C_S and the operational year of the unit, the net present value (NPV) is calculated. This, along with $C_{R\&M}$ and cash flow, helps analyse the life cycle cost analysis of the dryer system (Barnwal & Tiwari, 2008). Based on these factors, the payback period (C_N), the time required to recover investment cost and the capital recovery factor (CRF) are determined. The product of CRF and C_N gives the annualised uniform cost for a drying system. In the case of solar dehydration, the savings per drying units (S_{unit}) are determined taking into account the cost of the fresh produce (C_{fp}), cost of dried products (C_{dp}), cost of drying per unit of dried product (C_s) and comparing with the selling price of similar marketed product (C_b).

$$S_{unit} = \left(C_b - \left(\left(C_{fp} \times \frac{M_f}{M_d} \right) + C_s \right) \right)$$

where M' and M_d are the mass of fresh and dried products, respectively.

With this saving per unit in hand, the saving over the operation period (days/months/year) can be determined to calculate the NPV.

Based on the cash flow for the period, the internal rate of returns (IRR), benefit-to-cost ratio (BCR) and RoI are determined.

$$C_D = \frac{(C_I - C_S)}{N}$$

where N is the design life of a dryer (years) (typically considered 15 years)

$$NPV = \sum_{t=0}^{n} \frac{(A_t)}{(1+i)^t}$$

where A_t – cash flow for the period t or at a particular time and i is the annual minimal acceptable interest rate/discounted rate.

NPV depends on the cash flow and the interval of time between the cash flows. As seen in the equation, it is the value of all future cash flows over the entire life of an investment discount to the present.

$$NPV = \sum_{t=0}^{n} \frac{(A_t)}{(1+IRR)^t} = 0$$

With the NPV of all cash flows equated to zero, IRR is determined. It is used to determine the profitability of potential investments.

The payback period (C_N) is then calculated as:

$$C_N = \frac{C_I}{A_t}$$

BCR is another term for projecting the economic analysis. It projects the relationship between relative cost and the benefit of a proposed project, in qualitative or monetary terms.

$$BCR = \frac{\sum_{t=0}^{n} \frac{(B_t)}{(1+i)^t}}{\sum_{t=0}^{n} \frac{(C_t)}{(1+i)^t}}$$

where B_t is the benefit of the project in tth year, C_t is the cost of the project in tth year and i is the discount rate (Table 9.4).

$$RoI = \frac{Profit}{Total\,Capital\,Investment}$$

TABLE 9.4
Details of Economic Evaluation Parameters Considered in Techno-Economic Feasibility Studies

Sr. No.	Description	Notation	Unit	Remarks
C			**Economic Performance**	
1	Sizing/capacity of the dryer	C_{cap}		Defined based on the technical performance and operating performance
2	Area covered	A_{cap}	ha/year	
3	Material losses anticipated	M_{loss}	%	Accounts for the waste material that needs to be removed as a part of pre-processing and the material lost during the drying operation. It can be considered in the range of 2% to 10% of the fresh weight for the herbs, spices and medicinal plants

(Continued)

TABLE 9.4 (*Continued*)
Details of Economic Evaluation Parameters Considered in Techno-Economic Feasibility Studies

Sr. No.	Description	Notation	Unit	Remarks
4	Product output	M_{prod}	Tons	Dried product processed through the unit
5	Operation days	OD_{pm}	Per month	Typically considered to be 20–25 days
		OD_{py}	Per year	Typically considered to be 265–300 days depending on the harvest
6	Production	$C_{prod(pd)}$	Per day	Annual product output distributed over the operational periods
		$C_{prod(pm)}$	Per month	
		C_{prod}	Per year	
7	Initial cost for the dryer	C_I	USD	The cost of equipment includes the drying chamber, utilities and auxiliary units Installation and commissioning cost Ci&c is typically 10% of the equipment cost Hence, the initial cost of the dryer Ci = Ce + Ci&c
8	Design life	N	Years	Typically considered to be 5–15 years
9	Scrap value	C_S	USD	Can be considered from 0% to 25% of the initial cost
10	Depreciation cost	C_D	USD	It is the cost of actual cash value of the unit with the decided depreciation rate and design life
11	Repair and maintenance cost	$C_{R\&M}$	USD	It can be considered in the range of 2%–10% based on the type of dryers
12	Labour cost	C_{Lcost}	USD	Based on the wages, number of shifts and the number of workers required for the operation of the drying system, the annual cost associated with labour is to be determined
13	Utility requirement	C_U	USD	Total energy consumed for a selected type of dryer over the operational period
14	Total operating costs	C_{TO}	USD	Sum of utility cost and labour cost
15	Net raw material cost	C_{fresh} or C_{raw}	USD	The raw material purchase cost for the processing
16	Net dried product cost	C_{dried} or C_{prod}	USD	Typically, this is the sum of raw material cost, processing cost and profit margin associated distributed over the dried product quantity
17	Net earnings/net cash flow	A_t	USD	Difference of selling price of the product and raw material cost with processing cost

(*Continued*)

TABLE 9.4 (*Continued*)

Details of Economic Evaluation Parameters Considered in Techno-Economic Feasibility Studies

Sr. No.	Description	Notation	Unit	Remarks
	Discount rate considered	*I*	%	It is the key variable that determines the NPV of the investment. May vary from 7%, 10%, 14%, 21%, and up to 28%
18	Payback period for the dryer	C_N	Years	Time required to recover the investment
19	Net present value	*NPV*	USD	Provides the economic viability of the process Positive NPV indicates an economically viable investment or project, while a negative one shows that it is not financially feasible
20	Internal rate of returns	*IRR*	%	IRR is the discount rate that makes the NPV of all cash flows from a particular investment equal to zero Higher IRR indicates the more desirable project
21	Benefit–cost ratio	*BCR*		It is the ratio of total discounted benefits to total discounted costs. Project with a ratio greater than 1 has more significant benefits and is favourable

9.7 TECHNO-ECONOMIC EVALUATION – CASE STUDY

The mentioned method of TEFS can be elaborated for different dryers and products to provide ease of understanding. The first selected dryer is a typical hot air dryer powered by liquefied petroleum gas (LPG)-fired air heater of 33.3 kW. A forced draft is generated by the radial fan at a pressure of 250 kPa and 3,600 m³/h. The drying cabinet considered has dimensions of 4.0 m × 2.0 m × 1.0 m divided into four equal sections which can be mounted by four trays. It provided control over the inlet temperature using a thermostat (Soysal & Öztekin, 2001). Using this dryer, TEFS for dehydration of *M. piperita*, *Curcuma longa* and *Zingiber officinale* is projected herewith. It is recommended that low-priced raw materials like *M. piperita*, lemongrass, oregano and so on be dried by the cultivators only so that better benefits of the projects are seen.

Secondly, a greenhouse-type solar tunnel dryer with an approximate size of 10.0 m × 4.0 m × 2.0 m dimensions and semi-circular roof structure is selected. A provision for five chimneys and exhaust fans of 450 W capacity at the front and back is provided to ensure a draft to remove the moisture-laden air. No other additional utility is accounted. The material is loading about 150 kg per batch equally distributed over four arrays of 16 trays, each made up of SS wire mesh (Chavan & Thorat, 2021). Lastly, TEFS is projected for a solar conduction dryer (SCD), utilising solar energy in the form of conductive, convective and radiative heat transfer. The modular SCD consists of four drying chambers covered with a multi-walled polycarbonate sheet and fitted with drying trays of 1 m² each (Chavan & Thorat, 2021) (Tables 9.5–9.8).

TABLE 9.5
Description of Dryers (Chavan & Thorat, 2021; Soysal & Öztekin, 2001)

Sr. No.	Description	Conditions		
1	Dryer	Hot air dryer	Solar tunnel dryer	Solar conduction dryer
2	Mode of air flow	Radial fan forced draft	Radial fan and chimney-based draft	Natural convective draft
3	Loading capacity	130–150 kg/batch	150 kg/batch	150 kg/batch (10 units)
4	Drying area	27.36 m²	32 m²	40 m² (10 units)
5	Installation cost per m²	100	470	375
6	Drying efficiency	40%–60%	30%	45%–50%
7	Energy received	5,896 kJ/kg water evaporated	6.14×10^5 kJ of solar energy received	3.84×10^5 kJ of solar energy received

TABLE 9.6
Techno-Economic Evaluation for Drying Mentha, Turmeric and Ginger as Selected Crops in a Typical Hot Air Dryer (Soysal & Öztekin, 2001)

Sr. No.	Description	Notation	Unit	Crop Selected		
				Mentha	Turmeric	Ginger
A			Technical Performance			
1	Initial moisture content	M_i	%	80.01%	80.00%	80.00%
2	Desired moisture content	M_f	%	14.31%	5.00%	8.00%
3	Drying time	t_d	H	9	11	11
4	Loading capacity	B	Kg	145	100	100
5	Average drying rate	DR	kg/h	16.11	9.09	9.09
6	Water evaporate	M_w	kg	111.17	78.95	78.26
7	Moisture extraction rate	MER	kg/h	12.35	7.18	7.11
8	Air flow rate	M_a	kg/h	19,500	19,500	19,500
9	Specific heat of air	C_{Pair}	kJ/kg K	1	1	1
10	Inlet temperature	T_{Ai}	C	25	25	25
11	Outlet temperature	T_{Ao}	C	46	46	46

(*Continued*)

TABLE 9.6 (*Continued*)
Techno-Economic Evaluation for Drying Mentha, Turmeric and Ginger as Selected Crops in a Typical Hot Air Dryer (Soysal & Öztekin, 2001)

Sr. No.	Description	Notation	Unit	Crop Selected		
				Mentha	Turmeric	Ginger
	Latent heat of vaporisation	Λ	kJ/kg	2,260	2,260	2,260
12	Heat utilised to evaporate water	Q_e		660,753	587,921	586,370
13	Energy (utility) consumption	Q_h		1,180,509.96	1,034,497.00	1,013,157.60
	LPG	–	kg/kg db	32.079	27.875	27.300
	Efficiency	–	kg/h			
	Low heating value of LPG	–	kJ/kg			
14	Energy efficiency of the dryer	$E_{ta} = Q_e/Q_h$	%	55.97%	56.83%	57.88%
15	Specific heat energy requirement	Q_h	kJ	1,180,509.96	1,034,497.00	1,013,157.60
16	Power consumption (blower)	1	hp	0.745	0.745	0.745
17	Power factor of motor	0.8	f	0.80	0.80	0.80
18	Specific heat energy consumption	$Q_s = Q_h/M_w$	kJ/kg	10,618.58	13,103.63	12,945.90
19	Specific power consumption	$Q_p = fPt/M_w$	kJ/kg	173.70	298.95	301.58
20	Total energy requirement	$Q_t = Q_h + fPt$	kJ	1,180,683.66	1,034,795.95	1,013,459.18
B			**Operational Performance**			
1	Satisfactory annual harvest	Y	Tons/ha/year	37.5	22.5	12
2	No. of harvest	N	Per year	5	1	1
3	Yield per harvest	Y_i	Tons/ha	7.5	22.5	12
4	Harvest seasons	D	Days	14	45	45

(*Continued*)

TABLE 9.6 (*Continued*)
Techno-Economic Evaluation for Drying Mentha, Turmeric and Ginger as Selected Crops in a Typical Hot Air Dryer (Soysal & Öztekin, 2001)

Sr. No.	Description	Notation	Unit	Crop Selected		
				Mentha	Turmeric	Ginger
5	Peak processing requirement	L_{peak}	kg/day	0.54	0.50	0.27
6	Feasible number of batches	n_{batch}		2.00	2.00	2.00
7	Initial moisture content	M_i	%	80.01%	80.00%	80.00%
8	Final moisture content	M_f	%	14.31%	5.00%	8.00%
9	Drying time	t_d	h	9	11	11
10	Area per unit	A_u	m²	27.36	27.36	27.36
11	Loading density	L_d	kg/m²	5.300	3.655	3.655
12	Fresh biomass cost	C_{fresh}	USD/ kg	0.6	0.65	0.65
13	Dried biomass cost	C_{dried}	USD/ kg	2.2	2.5	2.8
C			**Economic Performance**			
1	Sizing/capacity of the dryer (batch)	C_{design}	kg	145	100	100
2	Area covered	A_{cap}	ha/year	1.30	2.13	4.00
3	Material losses anticipated	C_{loss}	%	1.00%	1.00%	1.00%
4	Material processed per year	$C_{pro(py)}$	Tons/ year	48.72	48	48
5	Evaporated water	M_w	kg/year	37,354.5	37,894.7	37,565.2
6	Throughput of dried biomass	$C_{dp(py)}$	kg/year	11,252	10,004	10,330
7	Operation days	OD_{pm}	Per month	14	20	20
		OD_{py}	Per year	168	240	240
8	Production	$C_{prod(pd)}$	Per day	66.975	41.684	43.043
		$C_{prod(pm)}$	Per month	937.657	833.684	860.870

(Continued)

TABLE 9.6 (*Continued*)
Techno-Economic Evaluation for Drying Mentha, Turmeric and Ginger as Selected Crops in a Typical Hot Air Dryer (Soysal & Öztekin, 2001)

Sr. No.	Description	Notation	Unit	Mentha	Turmeric	Ginger
				Crop Selected		
9	Initial cost for the dryer	C_i	USD	2,500	2,500	2,500
10	Design life	N	Years	15	15	15
11	Scrap value	C_s	USD	0	0	0
12	Depreciation cost	C_d	USD	166.67	166.67	166.67
13	Repair and maintenance cost	$C_{R\&M}$@5%	USD	125	125	125
14	Labour cost	C_L@20 USD/day	USD	3,360	4,800	4,800
a	Utility requirement	C_u				
b	Annual LPG consumption	$C_{U(LPG)}$	kg	16,710.96	12,000.0	13,440.0
b	Annual electricity consumption	$C_{U(elec)}$	kW	3,024	5,280	5,280
b	LPG	$C_{U(LPG)}$'@ USD 0.5872/kg	USD	9,812.68	7,046.40	7,891.97
16	Power	'@USD 0.0674/kWh	USD	203.82	355.87	355.87
17	Total operating costs	C_{op}		13,668.16	12,493.94	13,339.51
a	Net fresh product cost	C_{fresh}	USD	6,751.1	6,502.7	6,714.8
17b	Net dried product cost	C_{dried}	USD	24,754.14	25,010.53	28,925.22
18	Net earnings/net cash flow	At		11,085.98	12,516.59	15,585.71
19	Discount rate considered	I	%	10.00%	10.00%	10.00%
20	Payback period for the dryer	C_N	Years	0.577	0.416	0.282
21	Net present value	NPV	USD	9,516	14,050	29,048
22	Internal rate of returns	IRR		24%	25%	38%

TABLE 9.7

Techno-Economic Evaluation for Drying Mentha, Turmeric and Ginger as Selected Crops in a Typical Solar Tunnel Dryer

Sr. No.	Description	Notation	Unit	Crop Selected Mentha	Turmeric	Ginger
A			**Technical Performance**			
1	Initial moisture content	M_i	%	88.00%	82.00%	82.00%
2	Desired moisture content	M_f	%	14.00%	4.00%	7.50%
3	Drying time	t_d	H	6	8	8
4	Loading capacity	B	Kg	95	100	100
5	Average drying rate	DR	kg/h	15.83	12.50	12.50
6	Water evaporated	M_w	Kg	81.74	81.25	80.54
7	Moisture extraction rate	MER	kg/h	13.62	10.16	10.07
8	Latent heat of vaporisation	Λ	kJ/kg	2,260	2,260	2,260
9	Heat required to evaporate water	Q_e		184,742	183,625	182,022
10	Energy efficiency of the dryer	$E_{ta} = Q_e/Q_h$	%	30.00%	30.00%	30.00%
11	Energy (solar) absorption required	Q_h		615,806.20	612,083.33	606,738.74
12	Energy absorption per unit	Q_{abs}	kJ	614,000.00	614,000.00	614,000.00
13	Number of units based on receivable energy	$N_{(Unit)}$	Unit	1.00	1.00	1.00
14	Area availability per unit	A_{cap}	m²	32.00	32.00	32.00
15	Specific heat energy requirement	Q_h	kJ	614,000.00	614,000.00	614,000.00
16	Power consumption (blower)	0.6	hp	450	450	450
17	Power factor of motor	0.8	f	0.80	0.80	0.80
18	Specific heat energy consumption	$Q_s = Q_h/M_w$	kJ/kg	7,533.33	7,533.33	7,533.33
19	Specific power consumption	$Q_p = fPt/M_w$	kJ/kg	95,126.03	127,606.15	128,730.20
20	Total energy requirement	$Q_t = Q_h + fPt$	kJ	710,932.23	739,689.49	735,468.94
B			**Operational Performance**			
1	Satisfactory annual harvest	Y_a	Tons/ha/year	37.5	22.5	12
2	No. of harvest	n	Per year	5	1	1
3	Yield per harvest	Y_i	Tons/ha	7.5	22.5	12
4	Harvest seasons	D	Days	14	45	45
5	Peak processing requirement	$C_{pro(pd)}$	kg/day	0.54	0.50	0.27
6	Feasible number of batches			1.00	1.00	1.00

(Continued)

TABLE 9.7 (*Continued*)
Techno-Economic Evaluation for Drying Mentha, Turmeric and Ginger as Selected Crops in a Typical Solar Tunnel Dryer

Sr. No.	Description	Notation	Unit	Crop Selected		
				Mentha	Turmeric	Ginger
7	Initial moisture content	M_i	%	88.00%	82.00%	82.00%
8	Final moisture content	M_f	%	14.00%	4.00%	7.50%
9	Drying time	t_d	h	6	8	8
10	Area per unit	$A_{(unit)}$	m²	32.00	32.00	32.00
11	Loading density	L_D	kg/m²	2.969	3.125	3.125
12	Fresh biomass cost	C_{fresh}	USD/kg	0.6	0.65	0.65
13	Dried biomass cost	C_{dried}	USD/kg	2.2	2.5	2.8
C			**Economic Performance**			
1	Sizing/capacity of dryer (batch)	C_{design}	kg	95	100	100
2	Area covered	A_{cap}	ha/year	0.43	1.07	2.00
3	Material losses anticipated	C_{loss}	%	1.00%	1.00%	1.00%
4	Material processed per year	$C_{pro(py)}$	Tons/year	15.96	24	24
5	Evaporated water	M_w	kg/year	13,733.0	19,500.0	19,329.7
6	Throughput of dried biomass	$C_{dp(py)}$	kg/year	2,205	4,455	4,624
7	Operation days	OD_{pm}	Per month	14	20	20
		OD_{py}	Per year	168	240	240
8	Production	$C_{prod(pd)}$	Per day	13.123	18.563	19.265
		$C_{prod(pm)}$	Per month	183.726	371.250	385.297
9	Initial cost for the dryer	C_i	USD	10,025	10,025	10,025
10	Design life	N	Years	15	15	15
11	Scrap value	C_S	USD	0	0	0
12	Depreciation cost	C_d	USD	668.33	668.33	668.33
13	Repair and maintenance cost	$C_{R\&M}$@1%	USD	50.125	50.125	50.125
14	Labour cost	C_L@10 USD/day	USD	1,680	2,400	2,400
15	Utility requirement	C_U				
b	Annual electricity consumption	$C_{U(elec)}$	kW	1,008	1,920	1,920
b	Power	'@USD 0.0674/ kWh	USD	67.94	129.41	129.41
16	Total operating costs			2,466.40	3,247.87	3,247.87
17	Net dried product cost	'@USD/kg	USD	4,850.36	11,137.50	12,945.99

(*Continued*)

TABLE 9.7 (*Continued*)

Techno-Economic Evaluation for Drying Mentha, Turmeric and Ginger as Selected Crops in a Typical Solar Tunnel Dryer

Sr. No.	Description	Notation	Unit	Crop Selected		
				Mentha	Turmeric	Ginger
18	Net earnings/net cash flow	A_t		2,383.96	7,889.63	9,698.12
19	Discount rate considered	I	%	10.00%	10.00%	10.00%
20	Payback period for the dryer	C_N	Years	9.447	2.007	1.498
21	Net present value	NPV	USD	6,520	13,196	22,587
22	Internal rate of returns	IRR		3%	28%	40%

TABLE 9.8

Techno-Economic Evaluation for Drying Mentha, Turmeric and Ginger as Selected Crops in a Typical Solar Conduction Dryer

Sr. No.	Description	Notation	Unit	Crop Selected		
				Mentha	Turmeric	Ginger
A			**Technical Performance**			
1	Initial moisture content	M_i	%	88.00%	82.00%	82.00%
2	Desired moisture content	M_f	%	14.00%	4.00%	7.50%
3	Drying time	t_d	h	6	8	8
4	Loading capacity	B	kg	100	100	100
5	Average drying rate	DR	kg/h	16.67	12.50	12.50
6	Water evaporated	M_w	kg	86.05	81.25	80.54
7	Moisture extraction rate	MER	kg/h	14.34	10.16	10.07
8	Latent heat of vaporisation	Λ	kJ/kg	2,260	2,260	2,260
9	Heat required to evaporate water	Q_e		194,465	183,625	182,022
10	Energy efficiency of the dryer	$E_{ta} = Q_e/Q_h$	%	50.00%	50.00%	50.00%
11	Energy (solar) absorption required	Q_h		388,930.23	367,250.00	364,043.24
12	Energy absorption per unit	Q_{abs}	kJ	384,000.00	384,000.00	384,000.00
13	Number of units based on receivable energy	$N_{(Unit)}$	Unit	1.01	0.96	0.95

(*Continued*)

TABLE 9.8 (Continued)
Techno-Economic Evaluation for Drying Mentha, Turmeric and Ginger as Selected Crops in a Typical Solar Conduction Dryer

				Crop Selected		
Sr. No.	Description	Notation	Unit	Mentha	Turmeric	Ginger
14	Area availability per unit	A_{cap}	m²	4.00	4.00	4.00
15	Specific heat energy requirement	Q_h	kJ	384,000.00	384,000.00	384,000.00
18	Specific heat energy consumption	$Q_s = Q_h/M_w$	kJ/kg	4,520.00	4,520.00	4,520.00
B			**Operational Performance**			
1	Satisfactory annual harvest	Y_a	Tons/ha/year	37.5	22.5	12
2	No. of harvest	n	Per year	5	1	1
3	Yield per harvest	Y_i	Tons/ha	7.5	22.5	12
4	Harvest seasons	D	Days	14	45	45
5	Peak processing requirement	$C_{pro(pd)}$	kg/day	0.54	0.50	0.27
6	Feasible number of batches			1.00	1.00	1.00
7	Initial moisture content	M_i	%	88.00%	82.00%	82.00%
8	Final moisture content	M_f	%	14.00%	4.00%	7.50%
9	Drying time	t_d	h	6	8	8
10	Area per unit	$A_{(unit)}$	m²	4.00	4.00	4.00
11	Loading density	L_D	kg/m²	25.000	25.000	25.000

9.8   CONCLUDING REMARKS

Drying of herbs, spices and medicinal herbs is an important unit operation. The associated characteristics of each plant are owing to the volatile oils that get lost due to over-drying or high draft or temperature. Controlled drying is crucial in obtaining quality product to meet the vast market and consumer demand. The uniqueness of every herb, spice or medicinal plant and the difficult selection of the most suitable over-drying method makes the optimisation study for each of the products crucial with respect to its applications. Research advances have led to the development of many dryers for imparting the necessary energy of activation for water molecules. The energy is mainly required for the two sequentially and parallelly happening processes such as mobilising water molecules within the matrix and transferring molecules outside the matrix while preserving the desired characteristics. Selection of the most appropriate dryer for dehydration is equally important, hence the need for a techno-economic feasibility study.

TEFS was carried out considering the technical performance and economic evaluation of the dryer. In this chapter, the details of TEFS specific to the dehydration of herbs, spices and medicinal herbs were discussed. The need for TEFS, methodologies for conducting and parameters considered the same were elaborated. Guidelines for the determination of suitable dryer capacity, determination of area under cultivation and carrying out an economic evaluation accordingly were properly summarised using a case study for the hot air dryer, solar tunnel dryer and SCD for dehydration of mint leaves (*M. piperita*), turmeric (*Curcuma longa*) and ginger (*Zingiber officinale*).

REFERENCES

Babu, A. K., Kumaresan, G., Raj, V. A. A., & Velraj, R. (2018). Review of leaf drying: Mechanism and influencing parameters, drying methods, nutrient preservation, and mathematical models. *Renewable and Sustainable Energy Reviews, 90*(December 2016), 536–556. https://doi.org/10.1016/j.rser.2018.04.002.

Balasubramanian, S., Roselin, P., Singh, K. K., Zachariah, J., & Saxena, S. N. (2016). Postharvest processing and benefits of black pepper, coriander, cinnamon, fenugreek, and turmeric spices. *Critical Reviews in Food Science and Nutrition, 56*(10), 1585–1607. https://doi.org/10.1080/10408398.2012.759901.

Barnwal, P., & Tiwari, G. N. (2008). Life cycle cost analysis of a hybrid photovoltaic/thermal greenhouse dryer. *The Open Environmental Journal, 2*(1), 39–46. https://doi.org/10.217 4/1874233500802010039.

Benmoussa, H., Elfalleh, W., He, S., Romdhane, M., Benhamou, A., & Chawech, R. (2018). Microwave hydrodiffusion and gravity for rapid extraction of essential oil from Tunisian cumin (Cuminum cyminum L.) seeds: Optimization by response surface methodology. *Industrial Crops and Products, 124*(May), 633–642. https://doi.org/10.1016/j.indcrop.2018.08.036.

Bhaskara Rao, T. S. S., & Murugan, S. (2021). Solar drying of medicinal herbs: A review. *Solar Energy, 223*(June), 415–436. https://doi.org/10.1016/j.solener.2021.05.065.

Bousbia, N., Abert Vian, M., Ferhat, M. A., Petitcolas, E., Meklati, B. Y., & Chemat, F. (2009). Comparison of two isolation methods for essential oil from rosemary leaves: Hydrodistillation and microwave hydrodiffusion and gravity. *Food Chemistry, 114*(1), 355–362. https://doi.org/10.1016/j.foodchem.2008.09.106.

Calín-Sánchez, Á., Lipan, L., Cano-Lamadrid, M., Kharaghani, A., Masztalerz, K., Carbonell-Barrachina, Á. A., & Figiel, A. (2020). Comparison of traditional and novel drying techniques and its effect on quality of fruits, vegetables and aromatic herbs. *Foods, 9*(9). https://doi.org/10.3390/foods9091261.

Chavan, A., & Thorat, B. (2020). Mathematical analysis of solar conduction dryer using reaction engineering approach. *International Journal of Chemical Reactor Engineering, 18*(5–6). https://doi.org/10.1515/ijcre-2019-0220.

Chavan, A., & Thorat, B. (2021). Techno-economic comparison of selected solar dryers: A case study. *Drying Technology,* 1–11. https://doi.org/10.1080/07373937.2021.1919141.

Chavan, A., Vitankar, V., Mujumdar, A., & Thorat, B. (2020). Natural convection and direct type (NCDT) solar dryers: A review. *Drying Technology, 22*, 1–22. https://doi.org/10.10 80/07373937.2020.1753065.

Embuscado, M. E. (2015). Spices and herbs: Natural sources of antioxidants - A mini review. *Journal of Functional Foods, 18*, 811–819. https://doi.org/10.1016/j.jff.2015.03.005.

Fudholi, A., Sopian, K., Gabbasa, M., Bakhtyar, B., Yahya, M., Ruslan, M. H., & Mat, S. (2015). Techno-economic of solar drying systems with water based solar collectors in Malaysia: A review. *Renewable and Sustainable Energy Reviews, 51*, 809–820. https:// doi.org/10.1016/j.rser.2015.06.059.

Genskow, L. R. (1994). Dryer scale-up methodology for the process industries. *Drying Technology*, *12*(1–2), 47–58. https://doi.org/10.1080/07373939408959949.

Golmakani, M. T., & Moayyedi, M. (2015). Comparison of heat and mass transfer of different microwave-assisted extraction methods of essential oil from Citrus limon (Lisbon variety) peel. *Food Science and Nutrition*, *3*(6), 506–518. https://doi.org/10.1002/fsn3.240.

Guan, W., Li, S., Yan, R., Tang, S., & Quan, C. (2007). Comparison of essential oils of clove buds extracted with supercritical carbon dioxide and other three traditional extraction methods. *Food Chemistry*, *101*(4), 1558–1564. https://doi.org/10.1016/j.foodchem.2006.04.009.

Hammouda, F. M., Saleh, M. A., Abdel-Azim, N. S., Shams, K. A., Ismail, S. I., Shahat, A. A., & Saleh, I. A. (2013). Evaluation of the essential oil of Foeniculum vulgare mill (fennel) fruits extracted by three different extraction methods by GC/MS. *African Journal of Traditional, Complementary and Alternative Medicines*, *11*(2), 277–279. https://doi.org/10.4314/ajtcam.v11i2.8.

Hossain, M. B., Barry-Ryan, C., Martin-Diana, A. B., & Brunton, N. P. (2010). Effect of drying method on the antioxidant capacity of six Lamiaceae herbs. *Food Chemistry*, *123*(1), 85–91. https://doi.org/10.1016/j.foodchem.2010.04.003.

Jin, W., Mujumdar, A. S., Zhang, M., & Shi, W. (2018). Novel drying techniques for spices and herbs: A review. *Food Engineering Reviews*, *10*, 34–45. https://doi.org/10.1007/s12393-017-9165-7.

Kafshgary, S. K., Movagharnejad, K., & Aghili, F. (2014). Experimental study and empirical modeling of the drying of mint leaves in a fluid bed dryer. *The 8th International Chemical Engineering Congress & Exhibition (IChEC 2014))*, *8*(1), 24–27.

Kamaliroosta, L. (2012). Extraction of cinnamon essential oil and identification of its chemical compounds. *Journal of Medicinal Plants Research*, *6*(4), 609–614. https://doi.org/10.5897/jmpr11.1215.

Kamarulzaman, A., Hasanuzzaman, M., & Rahim, N. A. (2021). Global advancement of solar drying technologies and its future prospects: A review. *Solar Energy*, *221*(April), 559–582. https://doi.org/10.1016/j.solener.2021.04.056.

Khaing Hnin, K., Zhang, M., Mujumdar, A. S., & Zhu, Y. (2019). Emerging food drying technologies with energy-saving characteristics: A review. *Drying Technology*, *37*(12), 1465–1480. https://doi.org/10.1080/07373937.2018.1510417.

Khan, M. I. H., Wellard, R. M., Nagy, S. A., Joardder, M. U. H., & Karim, M. A. (2017). Experimental investigation of bound and free water transport process during drying of hygroscopic food material. *International Journal of Thermal Sciences*, *117*, 266–273. https://doi.org/10.1016/j.ijthermalsci.2017.04.006.

Kobos, P. H., Drennen, T. E., Outkin, A. S., Webb, E. K., Scott, M., & Wiryadinata, S. (2020). Techno-economic analysis: Best practices and assessment tools. In *Sandia Report* (Issue December). https://www.osti.gov/servlets/purl/1738878

Kürkçüoğlu, M., Koşar, M., & Başer, K. H. C. (2007). Comparison of microwave-assisted hydrodistillation and hydrodistillation methods for Pimpinella anisum L. *7th International Symposium*, *4*, 1–8.

Lawless, J. (1992). *The Encyclopedia of Essential Oils-Complete Guide for the Use of Aromatic Oils in Aromatherapy, Herbalism, Health and Well-Being*. Conari Press. Haper Collins.

Leal, P. F., Almeida, T. S., Prado, G. H. C., Prado, J. M., & Meireles, M. A. A. (2011). Extraction kinetics and anethole content of fennel (Foeniculum vulgare) and anise seed (Pimpinella anisum) extracts obtained by soxhlet, ultrasound, percolation, centrifugation, and steam distillation. *Separation Science and Technology*, *46*(11), 1848–1856. https://doi.org/10.1080/01496395.2011.572575.

Majumder, P., Sinha, A., Gupta, R., & Sablani, S. S. (2021). Drying of selected major spices: Characteristics and influencing parameters, drying technologies, quality retention and energy saving, and mathematical models. *Food and Bioprocess Technology, 14,* 1028–1054.

Manzan, A. C. C. M., Toniolo, F. S., Bredow, E., & Povh, N. P. (2003). Extraction of essential oil and pigments from Curcuma longa [L.] by steam distillation and extraction with volatile solvents. *Journal of Agricultural and Food Chemistry, 51*(23), 6802–6807. https://doi.org/10.1021/jf030161x.

Marongiu, B., Piras, A., & Porcedda, S. (2004). Comparative analysis of the oil and supercritical CO_2 extract of Elettaria cardamomum (L). Maton. *Journal of Agricultural and Food Chemistry, 52*(20), 6278–6282. https://doi.org/10.1021/jf034819i.

Mujumdar, A. S. (2014). Section I - Fundamental aspects. In A. S. Mujumdar (Ed.), *Handbook of Industrial Drying* (Forth edn, pp. 32–155). CRC Press Taylor & Francis Group, Boca Raton.

Obeng-Akrofi, G., Akowuah, J. O., Maier, D. E., & Addo, A. (2021). Techno-economic analysis of a crossflow column dryer for maize drying in Ghana. *Agriculture (Switzerland), 11*(6), 1–15. https://doi.org/10.3390/agriculture11060568.

Opara, E. I., & Chohan, M. (2014). Culinary herbs and spices: Their bioactive properties, the contribution of polyphenols and the challenges in deducing their true health benefits. *International Journal of Molecular Sciences, 15*(10), 19183–19202. https://doi.org/10.3390/ijms151019183.

Orphanides, A., Goulas, V., & Gekas, V. (2016). Drying technologies: Vehicle to high-quality herbs. *Food Engineering Reviews, 8*(2), 164–180. https://doi.org/10.1007/s12393-015-9128-9.

Orsat, V., Vijaya Raghavan, G. S., & Sosle, V. (2008). Adapting drying technologies for agri-food market development in India. *Drying Technology, 26*(11), 1355–1361. https://doi.org/10.1080/07373930802333452.

Pai, K. R., Sindhuja, V., Ramachandran, P. A., & Thorat, B. N. (2021). Mass transfer "regime" approach to drying. *Industrial and Engineering Chemistry Research, 60*(26), 9613–9623. https://doi.org/10.1021/acs.iecr.1c01680.

Patel, K. K., & Kar, A. (2012). Heat pump assisted drying of agricultural produce - An overview. *Journal of Food Science and Technology, 49*(2), 142–160. https://doi.org/10.1007/s13197-011-0334-z.

Pearson, A. M., & Gillett, T. A. (1996). Herbs, spices, and condiments. In Processed Meats. *Journal of AOAC International, 79*(1), 199–199. https://doi.org/10.1093/jaoac/79.1.199.

Pise, V., Shirkole, S., & Thorat, B. N. (2022). Visualization of oil cells and preservation during drying of betel leaf (piper betle) using hot-stage microscopy. *Drying Technology.* https://doi.org/10.1080/07373937.2022.2048848.

Prothon, F., Ahrne, L., & Sjoholm, I. (2003). Mechanisms and prevention of plant tissue collapse during dehdration - A critical review. *Critical Reviews in Food Science, 43*(4), 447–479.

Purohit, P., & Kandpal, T. C. (2005). Solar crop dryer for saving commercial fuels: A techno-economic evaluation. *International Journal of Ambient Energy, 26*(1), 3–12. https://doi.org/10.1080/01430750.2005.9674966.

Qiu, L., Zhang, M., Mujumdar, A. S., & Liu, Y. (2020). Recent developments in key processing techniques for oriental spices / herbs and condiments: A review. *Food Reviews International,* 1–21. https://doi.org/10.1080/87559129.2020.1839492.

Radoj, M., Pavkov, I., Kovacevic, D. B., Putnik, P., Wiktor, A., Stamenkovic, Z., Kešelj, K., & Gere, A. (2021). Effect of selected drying methods and emerging drying intensification technologies on the quality of dried fruit: A review. *Process, 9*(132), 21.

Rahman, M. M., Kumar, C., Joardder, M. U. H., & Karim, M. A. (2018). A micro-level transport model for plant-based food materials during drying. *Chemical Engineering Science*, *187*, 1–15. https://doi.org/10.1016/j.ces.2018.04.060.

Raina, A. P., Kumar, A., & Dutta, M. (2013). Chemical characterization of aroma compounds in essential oil isolated from "holy basil" (Ocimum tenuiflorum L.) grown in India. *Genetic Resources and Crop Evolution*, *60*(5), 1727–1735. https://doi.org/10.1007/s10722-013-9981-4.

Savitha, S., Chakraborty, S., & Thorat, B. N. (2022). Microstructural changes in blanched, dehydrated, and rehydrated onion. *Drying Technology*, *40*(12), 2550–2567. https://doi.org/10.1080/07373937.2022.2078347.

Sellami, I. H., Bettaieb, I., Bourgou, S., Dahmani, R., Limam, F., & Marzouk, B. (2012). Essential oil and aroma composition of leaves, stalks and roots of celery (Apium graveolens var. dulce) from Tunisia. *Journal of Essential Oil Research*, *24*(6), 513–521. https://doi.org/10.1080/10412905.2012.728093.

Shahi, A. K., Chandra, S., Dutt, P., Kaul, B. L., Tava, A., & Avato, P. (1999). Essential oil composition of Mentha x piperita L. from different environments of north India. *Flavour and Fragrance Journal*, *14*(1), 5–8. https://doi.org/10.1002/(SICI)1099-1026(199901/02)14:1<5::AID-FFJ768>3.0.CO;2-3.

Shao, Q., Huang, Y., Zhou, A., Guo, H., Zhang, A., & Wang, Y. (2014). Application of response surface methodology to optimise supercritical carbon dioxide extraction of volatile compounds from Crocus sativus. *Journal of the Science of Food and Agriculture*, *94*(7), 1430–1436. https://doi.org/10.1002/jsfa.6435.

Simoes, A., & Hidalgo, C. (2021). *Observatory of Economic Complexity (OEC) - The Economic Complexity Observatory: An Analytical Tool for Understanding the Dynamics of Economic Development*. Workshops at the Twenty-Fifth AAAI Conference on Artificial Intelligence. Papers from the 2011 AAAI Workshop (WS-11-17).

Sourmaghi, M. H. S., Kiaee, G., Golfakhrabadi, F., Jamalifar, H., & Khanavi, M. (2015). Comparison of essential oil composition and antimicrobial activity of Coriandrum sativum L. extracted by hydrodistillation and microwave-assisted hydrodistillation. *Journal of Food Science and Technology*, *52*(4), 2452–2457. https://doi.org/10.1007/s13197-014-1286-x.

Soysal, Y., & Öztekin, S. (2001). Technical and economic performance of a tray dryer for medicinal and aromatic plants. *Journal of Agricultural and Engineering Research*, *79*(1), 73–79. https://doi.org/10.1006/jaer.2000.0668.

Subramanian, M., & Taghizadeh-Hesary, F. (2021). Techno-economic feasibility study method in startup financing. In *Investment in Startup and Small Business Finances* (pp. 107–136). World Scientific. https://doi.org/10.1142/9789811235825_0004

Tanko, H., Carrier, D. J., Duan, L., & Clausen, E. (2005). Pre- and post-harvest processing of medicinal plants. *Plant Genetic Resources*, *3*(2), 304–313. https://doi.org/10.1079/pgr200569.

Taoukis, P. S., & Richardson, M. (2007). Priciples of intermediate moisture foods and related technology. In Gustavo V. Barbosa-Cánovas, Anthony J. Fontana Jr., Shelly J. Schmidt, and Theodore P. Labuza (Eds.), *Water Activity in Food - Fundamentals and Applications* (pp. 273–313). Blackwell Publishing. John Wiley & Sons, Inc.

Thamkaew, G., Sjöholm, I., & Galindo, F. G. (2021). A review of drying methods for improving the quality of dried herbs. *Critical Reviews in Food Science and Nutrition*, *61*(11), 1763–1786. https://doi.org/10.1080/10408398.2020.1765309.

Uthpala, T. G. G., Navaratne, S. B., & Thibbotuwawa, A. (2020). Review on low-temperature heat pump drying applications in food industry: Cooling with dehumidification drying method. *Journal of Food Process Engineering*, (May), 13. https://doi.org/10.1111/jfpe.13502.

Vázquez-Fresno, R., Rosana, A. R. R., Sajed, T., Onookome-Okome, T., Wishart, N. A., & Wishart, D. S. (2019). Herbs and spices - Biomarkers of intake based on human intervention studies - A systematic review. *Genes and Nutrition*, *14*(1), 1–27. https://doi.org/10.1186/s12263-019-0636-8.

Wang, D., Dai, J. W., Ju, H. Y., Xie, L., Xiao, H. W., Liu, Y. H., & Gao, Z. J. (2015). Drying kinetics of American ginseng slices in thin-layer air impingement dryer. *International Journal of Food Engineering*, *11*(5), 701–711. https://doi.org/10.1515/ijfe-2015-0002.

Xiao, H., & Mujumdar, A. S. (2015). Chapter 12: Impingement drying: Applications and future trends. In P. K. Nema, B. P. Kaur, and A. S. Mujumdar (Eds.), *Drying Technologies for Foods: Fundamentals & Applications* (Issue 2014, pp. 279–299). New India Publishing Agency.

Yamini, Y., Bahramifar, N., Sefidkon, F., Saharkhiz, M. J., & Salamifar, E. (2008). Extraction of essential oil from Pimpinella anisum using supercritical carbon dioxide and comparison with hydrodistillation. *Natural Product Research*, *22*(3), 212–218. https://doi.org/10.1080/14786410601130349.

10 Future Prospect and Global Market Demand for Dried Herbs, Spices and Medicinal Plants

Nikita S. Bhatkar, Vimal, and Shivanand S. Shirkole
Institute of Chemical Technology Mumbai

CONTENTS

10.1   INTRODUCTION

Herbs, spices and medicinal plants are regarded as the real treasure for any country in the global market, owing to the quantum of exchange and the revenue associated with it. The global import and export of some spices and herbs were reported to be 6.299 and 6.580 million tonnes, respectively in 2019 (Food and Agriculture Organization, 2010). This import quantity has increased by 7.04% from last year (i.e. 2019), highlighting the importance of these commodities in the international market.

Herbs are non-woody stem herbaceous plants majorly used in food for culinary purposes. The use of herbs in food can be traced back to ancient times and nowadays they have become the part and parcel of some cuisines. Herbs such as thyme, rosemary, lavender, coriander, basil and many more are desired to get a particular flavour and blend of taste in many cuisines. The use of herbs is not only restricted to food applications. The term herbs is used most of the time with spices, and together it forms a group with multiple functions in food including imparting flavour and aroma to the food, having preservative actions, acting as antioxidant agents and so on (Embuscado, 2015; Martínez-Graciá, González-Bermúdez, Cabellero-Valcárcel, Santaella-Pascual, & Frontela-Saseta, 2015). The most commonly traded herbs include thymes, coriander,

DOI: 10.1201/9781003269250-10

basil and so on. The medicinal properties of herbs in association with the natural product make them even more of a potential source for herb-based medicine development. As per Food and Agricultural Organization (FAO), it is estimated that about 80% of the population from developing nations is dependent on these medicines.

Spices are a group of vegetable products (including seeds, etc.), rich in essential oils and aromatic principles, and are mainly used as condiments available as a whole or in a crushed or powdered form. Spices always have a unique place in world trade. Spices have an excellent contribution to the food's nutritional value, flavour, colour and therapeutic properties. Spices also have medicinal values like antiseptic, antibacterial, antifungal, anticancer and anti-inflammatory properties. India is the largest producer, consumer and exporter of spices in the world producing 75 spices out of 109 varieties of spices listed by the International Organization for Standardization (ISO). Chilli holds the largest share in Indian spices export market (38%) and has a share of 50% among the global chilli export, whereas spices and Indian spice oils and oleoresins hold a 70% share in the global spices export. Overall, spices rank 22nd in principal commodities, 4th in agricultural commodities and 1st in horticultural commodities. Indian spices export market raised from 1.208 million tonnes (2019–20) with a value of INR 2,206,280.00 to 1.565 million tonnes (2020–21) with a value of INR 2,719,320.00. Spices are high-value agricultural products that have a high market value. The spices market can give an emerging nation a chance to further get into the globalized trade and take advantage of emerging opportunities.

Medicinal plants happen to be one of the most important assets to a country; for instance, India with more than 7,000 plant species recognized as medicinal plants generates a revenue of INR 3,211 crores by exporting 134,500 million tonnes of these commodities (National Medicinal Plant Board, Ministry of Ayush, Government of India). For the global scenario, it is reported that in total, 52,885 medicinal plants are used worldwide with 42,000 different spices (Schippmann, Leaman, & Cunningham, 2002). The surge in the consumer's perspective towards more authentic and traditional methods has considerably increased the demand for these medicinal plants in the global market. As per the reports, the global trade of herbal plants went from a value worth of 1.3 billion USD in 1996 to 120 billion USD in 2019 (Food_and_Agriculture_ Organization, 2010 and Press_Information_Bureau_Government_of_India_Ministry_ of_Commerce_and_Industry, 2019). The land under cultivation for these medicinal plants also went up from 153,000 to 827,000 hectares from 1991 to 2001 in China, the world's leading exporter. Medicinal plants find their use in nutraceutical, pharmaceutical and cosmetic applications; the increased demand for these commodities with therapeutic, curative and physiological benefits has paved a way for a substantial increase in the demand for them. For decades, Asia is the centre for exporting medicinal plants to the world, with China and India being the leaders in the list. Countries like Egypt and Morocco from Africa, Poland, Bulgaria and Albania from Europe and Chile and Peru from South Africa are important suppliers to the world.

Thus, it is needless to say that the demand for these medicinal plants, herbs and spices is immense and it is expected to grow in near future. Cumulatively, for the three commodities, China, India, Canada, United States and Germany are responsible for 60% of export quantity globally, and the United States, Germany, China, Japan and Singapore credit for 50% of worldwide import of these commodities (Nguyen, Duong, & Mentreddy, 2019). For instance, for spices like ginger, China is reported to be the maximum quantity exporter to the world from 1993 to 2020, while Japan imported the most quantity of ginger in the world. Figure 10.1 gives the pictorial representation of the quantities of ginger exported and imported by these countries to and from different parts of the world, respectively. The present chapter deals with the discussion on the present scenario, the expected future demand and supply of these medicinal plants, herbs and spices.

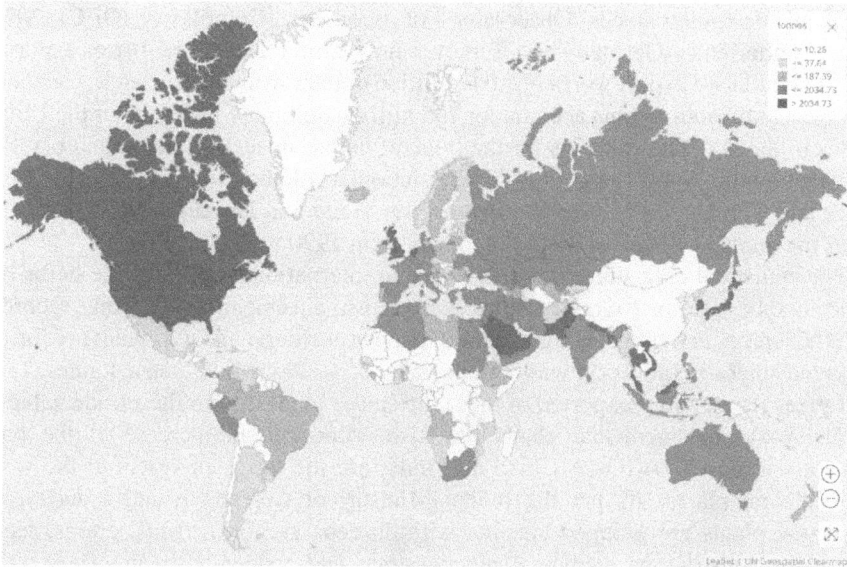

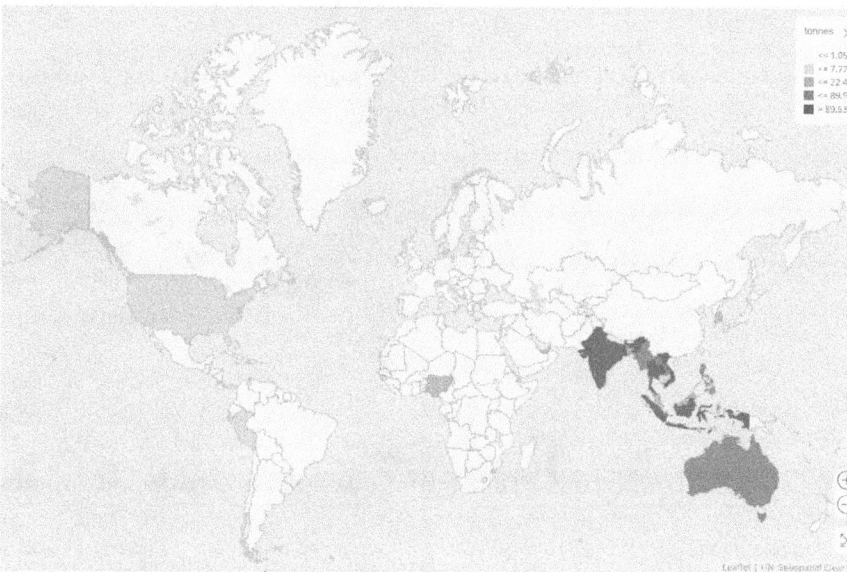

FIGURE 10.1 (a) Worldwide export quantity of ginger from China to the world from 1993 to 2020. (b) Import quantity of ginger from the world to Japan from 1993 to 2020. (Source: FAOSTAT.)

10.2 PRESENT GLOBAL MARKET SCENARIO FOR DRIED HERBS, SPICES AND MEDICINAL PLANTS

The advent of rise in different deadly infectious diseases like COVID-19 in this decade has led to a surge in demand for herbs, spices and medicinal plants dramatically. The annual quantum of spices traded in the world was accounted to be 3.1 billion dollars worth, with China being the leading exporter contributing to about 26.4% of the total export and the United States being the leading importer with an import percentage of 11.2% of the traded spices (Observatory_of_Economic_Complexity_(OEC), 2020). India, Netherlands, Germany and Turkey follow China next in the list of exporters with a cumulative export worth of 1,047 million dollars while Netherlands, Germany, Malaysia and Saudi Arabia account for 795 million dollars of import after the United States in the list. The statistics for the trade of herbs, spices and medicinal plants is provided by the World Integrated Trade Solution, a collaborative body of The World Bank and The United Nations Conference on Trade and Development. Figure 10.2 gives the quantity of spices traded in tonnes from 1990 to 2020.

The statistical data states that there was no international trade for the herbs like thyme and bay leaf in 2021. Nevertheless, there was a tremendous exchange recorded for other spices and herbs. Ginger, cumin, pepper, turmeric and coriander are the top exported spices in the world with a trade value of 1.728 million metric tonnes. Table 10.1 gives the quantity exported of these top spices in 2021 with their trade values.

The trade for medicinal plants has also witnessed an increase in the present years amidst COVID-19. China and India are the top 2 exporters to the world of medicinal plants. As per the Indian Ministry of Commerce and Industry, the medicinal plants are grouped together with oilseeds and olea, fruits, grains, seeds, and fruit industrial or medicinal plants, straw and fodder with an export value of INR 12,781.8 lakhs in 2022, with a marked increase of 32.65% from 2021

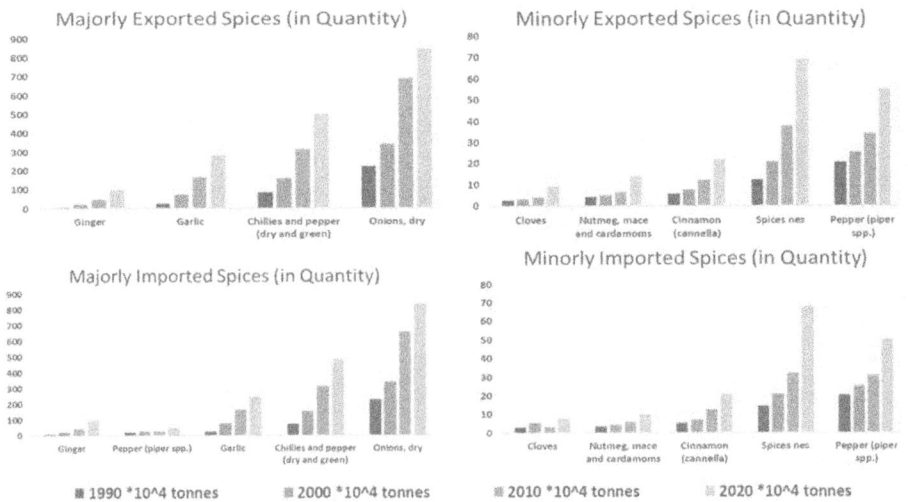

FIGURE 10.2 Quantity-wise import and export of spices for 1990, 2000, 2010 and 2020.

TABLE 10.1

Quantity-Wise Export of Top Spices in 2021 with Respective Trade Values

Spices	Quantity (kg)	Trade Value 1,000 USD
Ginger	820,266,247.66	1,117,517
Cumin seeds	279,879,396.81	597,990
Pepper (of the genus Piper), neither crushed nor ground	232,763,291.92	303,498
Turmeric (curcuma)	198,817,350.57	327,127
Coriander seeds	197,252,974.00	245,720
Cinnamon and cinnamon tree flowers, neither crushed nor ground	114,007,831.45	666,481
Anise or badian seeds	99,858,473.36	345,217
Pepper (of the genus Piper), crushed or ground	55,779,615.75	976,034
Cloves (whole fruit, cloves and stems)	41,761,076.93	275,750
Nutmeg	34,449,574.05	232,273
Cinnamon and cinnamon tree flowers, crushed or ground	31,593,575.56	171,525
Mace	6,148,767.291	102,383
Vanilla export	3,189,499	332,790
Saffron	1,715,553.736	128,414

(Ministry_of_Commerce_and_Industry_Govt._of_India, 2023). There are about 8,000 medicinal plant species in India, of which 90 are actively traded and among them, 178 are highly traded crossing 100 metric tonnes of quantity (Srirama et al., 2017). In the case of China, it has been reported that the export of herbal drugs is around 1.2 million tonnes which is way above those exported by India (Srirama et al., 2017).

10.3   PRESENT REGULATIONS FOR IMPORT/EXPORT OF DRIED HERBS, SPICES AND MEDICINAL PLANTS

The import and export throughout the world are majorly driven by the regulation imposed for the trade and exchange of a particular commodity. The regulatory bodies have a huge influence on the entire process of exchange for spices, herbs and medicinal plants. The standards are specific to the territory where the herbs, spices and medicinal plants are being imported, and the exporting countries are bound to follow them. It is worth noting that the regulations are not the same everywhere in the world; it varies from one country to another. As different limits and standards are set up by different countries like Canada, Australia and New Zealand, the maximum limit for aflatoxin is 15 ppb where in the United States, it is 20 ppb, which clearly defines the variation of standards among the countries. Presently, the European Spice Association, Australian Food and Grocery Council, American Spice Trade

Association (ASTA), Canadian Spice Trade Association, Netherlands Spice Trade Association, Danish Spices Association, Italian Association of the Food Products Industry, All Nippon Spices Association (Japan), AFEXPO (Spain), Seasoning and Spice Association (UK) and Livsmedelsforetagen (Sweden) are the Spice/Agricultural Trade Associations of the respective countries. As far as India is concerned, the Spice Board of India under the Ministry of Commerce and Industry looks into the affairs of the spice trade. Similarly, Administrative Measures for Food Trading Permits (2017) under China Food and Drug Administration is the concerned authority. As per the European Union regulations, pathogens like *Salmonella and E. coli* should be absent in the spices and herbs and *Enterobacteriaceae* and *Bacillus cereus*, yeast and mould should be less than 100 cfu/g. Limits for mycotoxins are specific to the product. Dried chillies and paprika, pepper, nutmeg, ginger and turmeric should be less than 5 µg/kg for aflatoxin B1 and 10 µg/kg for the sum of B1, B2, G1 and G2. Ochratoxin A should be less than 20 µg/kg for dried chillies and should be less than 15 µg/kg for pepper, nutmeg, ginger and turmeric (CBI_Ministry_of_Foreign_Affairs, 2023). Also, for irradiated spices and herbs, it is mandatory to mention the treatment on the label. Another possible hazard that has caught wide attention among consumers, researchers and industries is the presence of pesticide residues; the European Union provides a list of pesticides with the maximum residual limit in the European Union Directive on maximum residue levels on pesticides, and there are amendments done as and when required (Regulation_of_the_European_Parliament_and_of_the_Council, 2005). The Spice Board of India has provided the maximum limits for some parameters for different toxins and hazards for different countries; Table 10.2 summarizes some of them. The additions of adulterants in some cases can also lead to the indirect presence of some pesticide residues as observed for oregano which is often adulterated with leaves originating from olive trees, myrtle, sumac, cistus and

TABLE 10.2
Parameter and Maximum Limit for Certain Spices for Some Countries

Spice	Country	Parameters	Maximum Limit
Chilli whole	European Union (EU), Northern Ireland, United Kingdom, South Africa, Japan and Malaysia	Aflatoxin total	10 ppb
	Canada, Australia and New Zealand		15 ppb
	United States	Aflatoxin total	20 ppb
	China, Southeast Asian countries (excluding Malaysia), Sri Lanka and all other countries	Aflatoxin total	30 ppb
Turmeric whole	European Union (EU) and Northern Ireland United Kingdom	Aflatoxin total	10 ppb
Turmeric powder	Japan, North American countries (other than the United States), EU, United Kingdom, Australia and New Zealand	Sudan I–IV	Not detected

(Continued)

TABLE 10.2 *(Continued)*
Parameter and Maximum Limit for Certain Spices for Some Countries

Spice	Country	Parameters	Maximum Limit
Ginger	European Union (EU), Northern Ireland and United Kingdom	Aflatoxin total	10 ppb
	European Union (EU), Northern Ireland and United Kingdom	Ethylene oxide	0.02 ppm
Cumin Seeds	European Union (EU), Northern Ireland, United Kingdom, United States, South Africa, Japan, Bangladesh, Bhutan, Nepal, Pakistan, Afghanistan, China, Australia, New Zealand and all other countries	Extraneous matter	3.00% max.
Curry powder and masalas	China, Japan, Bangladesh, Australia, New Zealand, Bhutan, Nepal, United Kingdom, North American countries, South Africa, EU, Northern Ireland (UK), Pakistan, Afghanistan and all other countries	Sudan I–IV	Not detected
	Malaysia, Japan, South Africa, United Kingdom, European Union (EU) and Northern Ireland (UK)	Aflatoxin total	10 ppb
	Australia, New Zealand and Canada	Aflatoxin total	15 ppb
	United States	Aflatoxin total	20 ppb
	China, Bangladesh, Bhutan, Nepal, Pakistan, Afghanistan, North American countries (except the United States and Canada) and all other countries	Aflatoxin total	30 ppb

Source: Spice Board of India.

hazelnut which can potentially contain some pesticide residues. It was reported that of 76 samples, 34 were adulterated samples and the rest were genuine samples, and 55 different pesticide residues were identified from them (Drabova et al., 2019).

According to the Code of Federal Regulations report published in 2021, most herbs and spices are grouped together and named herbs and spices. The residual specifications related to tolerances of residues after fumigation and other antimicrobials used to decrease the microbial load for herbs and spices are provided. The tolerance of residues for propylene oxide is 300 ppm, and for propylene chlorohydrin, it is 1,500 ppm whereas for fluorides, sulphuryl fluoride, and carfentrazone-ethyl (which are mostly residues of herbicides), the limits are 70, 0.5 and 2.0 ppm, respectively. For peroxyacetic acid, the limit is 100 ppm which is used for antimicrobial treatment (Code_of_Fedral_Regulations_(CFR), 2021). Table 10.3 provides the tolerance limit for specific commodities.

TABLE 10.3

Tolerance Limit for Residues for Specific Commodities

Sr. No.	Residues	Tolerances for Residues (ppm)	Applicable for Commodity
1	Propylene oxide	300	Herbs and spices, group 19, dried post-harvest fumigant
2	Propylene chlorohydrin	1,500	Herbs and spices, group 19, dried, except basil post-harvest fumigant
3	Fluorides	70	Post-harvest fumigant
4	Sulphuryl fluoride	0.5	Herbs and spices, group 19, post-harvest fumigant
5	Carfentrazone-ethyl	2.0	Residues of herbicides
6	Peroxyacetic acid	100	Antimicrobial treatment

Source: CFR 2021.

10.4 ISSUES ASSOCIATED WITH THE GLOBAL MARKET DEMAND

The market for herbs, spices and medicinal plants is highly dynamic when it comes to the demand for these commodities. Various malpractices and adulterations in the commodities are conducted which affect the overall trade. The wide bridge between the supply and demand of spices, herbs and medicinal plants, the prevalent standards, the non-compliances by the exporting countries and unfair trade practices are the major issues in the global market.

The increasing demand for herbs, spices and medicinal plants can only be curbed by a continuous supply of these commodities by the exporting countries. However, the substantially lower productivity of these commodities is a major concern that the leading countries are facing. Leading nations like India have put forth this issue of lower productivity in terms of yield obtained per unit area under cultivation (Sharangi & Pandit, 2018). The country recorded the current yield of spices like black pepper and cardamom to be 283 and 181 kg/ha, and the demand is expected to increase by 20%–25% in the Western markets every year. Thus, the issue of low productivity is of high concern to be addressed. The use of chemical fertilizers, high labour cost, lack of resistant variety, crop diseases, loss of soil fertility and post-harvest losses are the various reasons for low productivity. The supply of spices and herbs in some countries is also observed to reduce due to the devastation of crops by some diseases or others. Ethiopia witnessed a reduction in the export of ginger as the crop was devastated due to the bacterial wilt epidemic in 2013 (Shimelis, 2021). Ginger was the highest exported spice in Ethiopia during 2007–2013; however, the export of ginger completely vanished due to the wilt epidemic in 2015. On the other hand, the other ruling country for the export of spices, herbs and medicinal plants, China, is facing issues related to the extinction of species, due to the monotonous growth of commonly traded herbs and spices (Leung, 2006). Two hundred species of plants have already become extinct, with 1,000 others under imminent threat due to over-exploitation. The process of soil remediation and the development of resistant variety can be practised to overcome these problems. Thus, the use of resistant variety and good agricultural practices at the farm level to get rid of pests and weeds is required.

The differences in the standards prevailing in different countries also lead to issues with some commodities. This difference in the standards is also seen in the case of herbs like oregano. The *Origanum* genus includes the botanical genera from both the Mediterranean and Mexico. Some specific genera like *Origanum vulgare L. ssp. hirtum* and *Origanum onites L* are mostly considered true genera with specifications like the level of impurities (extraneous material max. 2%) in the European market. Leaves of all *Origanum* genera are allowed in other markets with limitations specified in ISO/FDIS 7925 and ASTA guidelines. This in turn hampers the exporter country to some extent, and uniformity of the standards needs to be established throughout the globe. Nevertheless, Codex Alimentarius is trying continuously to harmonize the standards globally. Different countries are continuously setting different standards for particular spices and herbs like dried chillies, ginger and garlic given by India, coriander and basils by Egypt, cloves by Nigeria and nutmeg by Indonesia (Sharangi & Pandit, 2018).

Microbial contamination of food is a problem throughout the world and the section of trade of spices, herbs and medicinal plants is not unbarred by this problem. As per a survey conducted of the imported shipments to the United States, the imported shipments showed a higher prevalence of *Salmonella* than that of retail samples (Zhang et al., 2017). Another common issue with the trade of spices, herbs and medicinal plants is the malpractices carried out. Spices and herbs are the commodities with the highest number of fraud cases registered at the international level from 2000 to 2020, as per reports of Rapid Alert System for Food and Feed (RASFF). Figure 10.3 provides the pictorial representation of the same. The reports suggest issues related to adulterations, the presence of unwanted allergens, adherence of foreign materials, contaminations of microbial species and toxins are the most prevalent. The issues related to non-compliances are many; as per a survey conducted by RASFF, it was found 4,118 total samples to be non-compliant, and as per an inspection conducted in 2019, 110 samples were rejected for the spice and herb category mostly due to contaminations with *Salmonella* and *Clostridium botulinum*. Other unauthorized substances such as mycotoxins and pyrrolizidine alkaloids were the prominent ones in chillies.

Adulteration of spices and herbs is a common practice to gain economic benefits. European Union recently carried out an analysis of 1,000 samples of six different spices and herbs. It was reported that 48% of oregano, 17% of pepper, 14% of cumin, 11% of saffron and 6% of paprika were found to be at risk of adulteration (European_Commission, 2021). Sudan red is the most common adulterant in herbs and spices – as it imparts a bright colour to the spice mixes and powders – a carcinogenic compound. The scandal of Sudan dye in spices in 2004 cost around US$418 million. Sudan 1 was discovered to be illegally present in chilli powder and the meals containing chilli powder in the European Union in May 2003. Cumin is found to be adulterated with the shells of peanuts, which is very deleterious to allergic populations. Similarly, black pepper, turmeric and Chinese star anise are commonly adulterated with papaya seeds, lead chromate, and Japanese star anise, respectively.

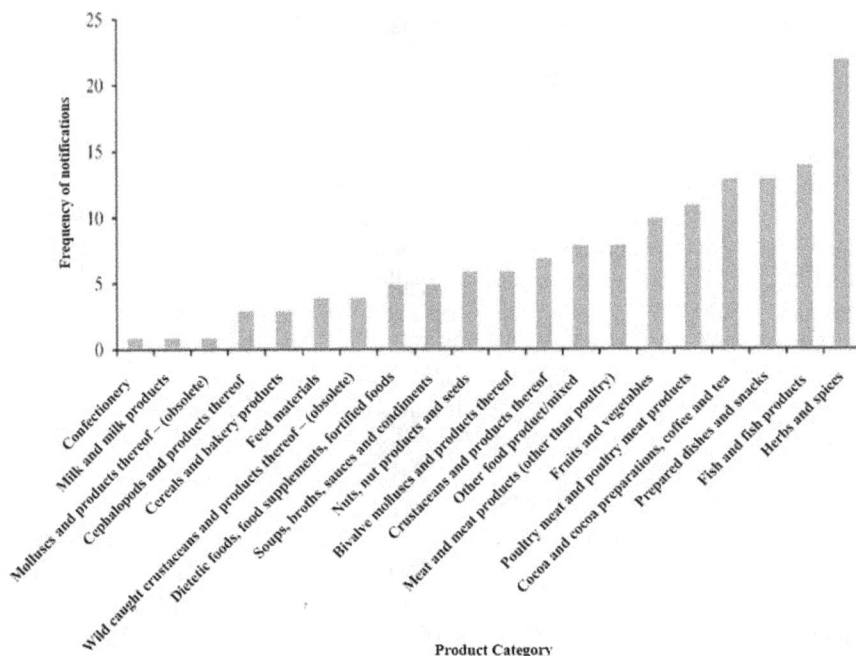

FIGURE 10.3 Frequency of notification for fraudulence in different food commodities during 2000–2020. (Source: Rapid alert system for food and feed, http://dx.doi.org/10.1371/journal.pone.0259298.)

For medicinal plants and herbal drugs, adulteration of species is commonly done, and until recent times, this issue was unknown. The extent of adulteration can be as high as 80% due to the less focus on the identification of plant species (Srirama et al., 2017). Recently, techniques such as mass spectroscopy, nuclear magnetic resonance, microscopy and DNA barcoding are helping in several ways to overcome this problem. These issues can only be solved by establishing the standards for the trade of herbal medicinal plants and doing more research in analytical techniques.

Other common examples of fraudulence include deliberate substitution, addition, tampering, misrepresentation and so on. Analytical tools such as sensory and organoleptic evaluations, micro- and macroscopy, DNA barcoding and various different spectroscopic methods can be useful to tackle such issues in future (Osman et al., 2019).

10.5   FUTURE PROSPECT AND GLOBAL MARKET DEMAND/CHALLENGES

The demand for spices and herbs will be expected to grow from US$4 billion to US$6.5 billion in near future. For dried herbs and spices, the three main markets in the European Union are retail, catering and food manufacturing (which is the highest among all three and contributes 50%–60% of trade). US consumers are developing an interest in the consumption of spices and herbs where most of these are consumed in Asia and Europe. With the irreplaceable role of spices, herbs and medicinal plants throughout the world for various nutraceutical, food preparation and medicinal purposes, it is needless to say that the demand for these commodities is going to touch the sky. As per the reports of the Spice Board of India, the demand for these is supposed to increase by many folds; to be precise, it is expected that the demand for black pepper, cardamom, turmeric and ginger is going to rise by 2–5 folds more than the current market demands at the international levels (Indian_Council_of_Agricultural_Research_(ICAR), 2013). The report also suggests that the demand for these commodities is way above (3.19%) the population growth rate in the world. To cater to this demand, there are several challenges to be tackled in near future. Continuous lowering of productivity in terms of area under cultivation is a challenging issue with the least control. One of the leading countries like India has put forth low productivity as the challenge to remain in the market with other competing countries like China. For instance, India reported the productivity of black pepper to be 283 kg/ha which is several times lower than a country like Malaysia (2,925 kg/ha). The issues related to unfair trade practices and the inability to meet the required standards are some of the important issues discussed in the article previously. The established standards as well as the standards to be established in future will play a key role in this context. The envisaged increase in the demand for herbs, spices and medicinal plants along with the issues discussed above puts forth the need for higher productivity, achievable standards, obeying international standards and fair trade to be focussed on to have a harmonized supply to serve the future demands and challenges.

10.6   CONCLUDING REMARKS

The demand for dried herbs, spices and medicinal plants in the global market is tremendous and is expected to grow gradually in future. Countries like China and India are the leading exporters in the world. The regulatory bodies play a vital role in establishing the standards and are responsible for carrying out fair trade at an international level. Issues such as adulterations, substitutions and other malpractices are still persistent at the global level, and appropriate analytical methods and stringent rules and regulations need to be implemented globally. The issues such as low productivity and extinction of species due to overexploitation need to be addressed judicially to have a sustainable approach throughout the supply chain of these commodities.

REFERENCES

CBI_Ministry_of_Foreign_Affairs. (2023). Exporting Spices and Herbs to Europe. Retrieved from https://www.cbi.eu/market-information/spices-herbs, Retrieved on November 15, 2022.

Code_of_Fedral_Regulations_(CFR). (2021). Tolerances and Exemptions for Pesticide Chemical Residues in Food. Retrieved from https://www.govinfo.gov/content/pkg/CFR-2021-title40-vol26/pdf/CFR-2021-title40-vol26-part180.pdf, Retrieved on September 18, 2022.

Drabova, L., Alvarez-Rivera, G., Suchanova, M., Schusterova, D., Pulkrabova, J., Tomaniova, M., … Hajslova, J. (2019). Food fraud in oregano: Pesticide residues as adulteration markers. *Food Chemistry*, *276*, 726–734. doi:10.1016/j.foodchem.2018.09.143.

Embuscado, M. E. (2015). Spices and herbs: Natural sources of antioxidants–a mini review. *Journal of Functional Foods*, *18*, 811–819. doi:10.1016/j.jff.2015.03.005.

European_Commission. (2021). Herbs and Spices (2019–2021). Retrieved from https://food. ec.europa.eu/safety/agri-food-fraud/eu-coordinated-actions/coordinated-control-plans/ herbs-and-spices-2019-2021_en. Retrieved on October 10, 2022.

Food_and_Agriculture_Organization. (2010). Sustainable Use of Medicinal Plants - A Multi Sectoral Challenge and Opportunity. https://www.fao.org/3/y4496e/Y4496E42.htm. Retrieved on December 16, 2022.

Indian_Council_of_Agricultural_Research (ICAR). (2013). Vision 2050. Retrieved from http://spices.res.in/sites/default/files/Vision-IISR-2050.pdf. Retrieved on December 10, 2022.

Leung, P.-C. (2006). Good Agricultural Practice—GAP Does It Ensure a Perfect Supply of Medicinal Herbs for Research and Drug Development? In *Current Review of Chinese Medicine: Quality Control of Herbs and Herbal Material* (pp. 27–57). World Scientific.

Martínez-Graciá, C., González-Bermúdez, C. A., Cabellero-Valcárcel, A. M., Santaella-Pascual, M., & Frontela-Saseta, C. (2015). Use of herbs and spices for food preservation: Advantages and limitations. *Current Opinion in Food Science*, *6*, 38–43. doi:10.1016/j. cofs.2015.11.011.

Ministry_of_Commerce_and_Industry_Govt._of_India. (2023). Export Import Data Bank Version 7.1-Tradestat. Retrieved from https://tradestat.commerce.gov.in/eidb/default. asp. Retrieved on January 10, 2023

Nguyen, L., Duong, L. T., & Mentreddy, R. S. (2019). The US import demand for spices and herbs by differentiated sources. *Journal of Applied Research on Medicinal Aromatic Plants*, *12*, 13–20. doi:10.1016/j.jarmap.2018.12.001.

Observatory_of_Economic_Complexity_(OEC). (2020). Latest Trade data of Spices. Retrieved from https://oec.world/en/profile/hs/spices#:~:text=In%202020%2C%20Spices%20 were%20the, 0.022%25%20of%20total%20world%20trade. Retrieved on September 25, 2022.

Osman, A. G., Raman, V., Haider, S., Ali, Z., Chittiboyina, A. G., & Khan, I. A. (2019). Overview of analytical tools for the identification of adulterants in commonly traded herbs and spices. *Journal of AOAC International*, *102*(2), 376–385. doi:10.5740/ jaoacint.18-0389.

Press_Information_Bureau_Government_of_India_Ministry_of_Commerce_and_Industry. (2019). Export of Herbs and Herbal Products. Retrieved on October 17, 2022.

Regulation_of_the_European_Parliament_and_of_the_Council. (2005). Maximum Residue Levels of Pesticides in Or on Food and Feed of Plant and Animal Origin. Retrieved from Official Journal of European Union. https://eur-lex.europa.eu/LexUriServ/LexUriServ. do?uri=OJ:L:2005:070:0001:0016:en:PDF. Retrieved on November 15, 2022.

Schippmann, U., Leaman, D. J., & Cunningham, A. (2002). Impact of Cultivation and Gathering of Medicinal Plants on Biodiversity: Global Trends and Issues. In *Biodiversity the Ecosystem Approach in Agriculture, Forestry Fisheries*.

Sharangi, A. B., & Pandit, M. (2018). Supply Chain and Marketing of Spices. In Sharangi, A. (ed.), *Indian Spices* (pp. 341–357). Springer.

Shimelis, T. (2021). Spices production and marketing in Ethiopia: A review. *Cogent Food Agriculture*, *7*(1), 1915558. doi:10.1080/23311932.2021.1915558.

Srirama, R., Santhosh Kumar, J., Seethapathy, G., Newmaster, S. G., Ragupathy, S., Ganeshaiah, K., ... Ravikanth, G. (2017). Species adulteration in the herbal trade: Causes, consequences and mitigation. *Drug Safety*, *40*(8), 651–661. doi:10.1007/s40264-017-0527-0.

Zhang, G., Hu, L., Pouillot, R., Tatavarthy, A., Doren, J. M. V., Kleinmeier, D., ... Brown, E. W. (2017). Prevalence of Salmonella in 11 spices offered for sale from retail establishments and in imported shipments offered for entry to the United States. *Journal of Food Protection*, *80*(11), 1791–1805. doi:10.4315/0362-028X.JFP-17-072.

Index

Note: **Bold** page numbers refer to tables and *italic* page numbers refer to figures.

For Product Safety Concerns and Information please contact our EU
representative GPSR@taylorandfrancis.com
Taylor & Francis Verlag GmbH, Kaufingerstraße 24, 80331 München, Germany

www.ingramcontent.com/pod-product-compliance
Lightning Source LLC
Chambersburg PA
CBHW060357220326
41598CB00023B/2947

9 781032 216188